PAPUA NEW GUINEA
PORT MORESBY
SOLOMON SEA
Gulf of Papua
Torres Strait
Cape York
Cape Wessel
Wessel Islands
Cape Wilberforce
Cape Arnhem
Cape Grey
Cape Shield
Groote Eylandt
Cape Beatrice
Limmen Bight
Gulf of Carpentaria
Sir Edward Pellew Group
Borroloola
Wellesley Islands
Duifken Point
Weipa
Pera Head
Cape York Peninsula
Cape Grenville
Cape Weymouth
Cape Direction
Iron Range
Great Barrier Reef
CORAL SEA
Osprey Reef
Cape Flattery
Cooktown
C. Tribulation
Cairns
Cape Grafton
Mareeba
Innisfail
Tully
Holmes Reef
Flora Reef
Herald Cays
Willis Group
Magdelaine Cays
Coringa Is.
CORAL SEA ISLANDS TERRITORY
Flinders Reefs
Diamond Is.
Tregosse Islets and Reefs
Lihou Reef and Cays
Burketown
Normanton
Croydon
Barkly Tableland
Gregory
Forsayth
Ingham
Hinchinbrook Island
Greenvale
Halifax Bay
Townsville
Ayr
Marion Reef
Tennant Creek
Charters Towers
Bowen
Whitsunday Group
Mount Isa
Cloncurry
Selwyn Ra.
Hughenden
Great Dividing Range
Mt Dalrymple
Mackay
Northumberland Isles
Swain Reefs
Frederick Reef
Ken Reef
Saumarez Reef
Winton
Boulia
QUEENSLAND
Wreck Reef
Capricorn Channel
Capricorn Group
TROPIC OF CAPRICORN
Longreach
Barcaldine
Rockhampton
Curtis I.
Gladstone
Cato Island and Bank
Blackall
Yaraka
Simpson Desert
Buckland Tableland
Bilpa Morea Claypan
Birdsville
Hervey Bay
Sandy Cape
Fraser I.
Bundaberg
Maryborough
Sturt Stony Desert
Quilpie
Charleville
Roma
Double Island Point
Gympie
Tewantin
Nambour
Maroochydore
Caboolture
Bongaree
Moreton Island
Grey Range
Cunnamulla
Beardmore Res.
Toowoomba
Brisbane
Ipswich
Lake Eyre North
Warwick
Gold Coast
Coober Peddy
Lake Gregory
Bulloo Lake
Lake Blanche
Lismore
Ballina
Lake Eyre South
Marree
Lake Callabonna
Narran Lake
Moree
TASMAN SEA
USTRALIA
Leigh Creek
Lake Frome
Bourke
Walgett
Grafton
Tarcoola
Lake Torrens
Lake Harris
Woomera
Round Mt
Coffs Harbour
Lake Gairdner
Flinders Ranges
Wilcannia
Armidale
Lake Macfarlane
Port Augusta
Broken Hill
Port Macquarie
Menindee Lake
NEW SOUTH WALES
Mt Barrington
Taree
Tandou Lake
Dubbo
Burrendong Res.
Muswellbrook
Forster-Tuncurry
Whyalla
Singleton
Port Pirie
Lake Victoria
Garnpung Lake
Parkes
Newcastle
Wyong
Bathurst
Penrith
Sydney
Spencer Gulf
Mildura
Griffith
Campbelltown
Wollongong
Port Lincoln
Cape Carnot
Cape Catastrophe
Adelaide
Crafers
Murray Bridge
Yorke Peninsula
Gulf St Vincent
Investigator Strait
Kangaroo Island
Goulburn
Kiama
Nowra
Lake Tyrrell
Wagga Wagga
CANBERRA
Albury
Echuca
Wodonga
Shepparton
Wangaratta
Horsham
VICTORIA
2230 Mount Kosciuszko
Cape Jaffa
Rocklands Res.
Ballarat
Melbourne
Geelong
Mount Gambier
Discovery Bay
Cape Nelson
Portland
Warrnambool
Colac
Frankston
Morwell
Bairnsdale
Sale
Gippsland
Cape Howe
Rame Head
Cape Otway
Wilson's Promontory
TASMAN SEA
Melbourne
Geelong
Frankston
Bass Strait
King Island
Furneaux Group
North Point
Hunter Islands
Cape Grim
Tasmania
Burnie
Devonport
Launceston
Eddystone Point
1617 Mount Ossa
TASMANIA
Hobart
Maria Island
South West Cape
Southeast Cape
TASMAN SEA

Terralog

herpetological reference of the world

Holger Vetter

Turtles of the World Vol. 5 Australia and Oceania

Schildkröten der Welt Band 5 Australien und Ozeanien

Edition Chimaira

Vetter, Holger:
TERRALOG:
Turtles of the World Vol. 5
Australia and Oceania
Schildkröten der Welt Band 5
Australien und Ozeanien
Frankfurt am Main
Edition Chimaira

ISBN 978-3-930612-99-4

Umschlagvorderseite:
Front cover: large photo/großes Foto:
Emydura subglobosa, captive female. J. Harding

smaller photo to the right/kleineres Foto rechts:
Elseya (Pelocomastes) irwini, Broken River, northeastern QLD, Australia. G. Schmida

smaller photo to the left/kleines Foto links:
Chelodina (Macrochelodina) oblonga oblonga, Darwin, NT, Australia. G. Schmida

large photo opposite to the symbol explanations/
großes Foto gegenüber der Tafel mit den Symbolerklärungen:
Chelodina (Macrochelodina) oblonga oblonga, Kakadu NP, NT, Australia.
B. Wright/Lochman LT

Heddernheimer Landstr. 20
60439 Frankfurt am Main
Germany
Phone: +49-69-497223
Fax: +49-69-497826
E-Mail: frogbook@aol.com
Internet: www.chimaira.de

Verleger/Publisher: Andreas S. Brahm
Satz/Layout: Dr. Gerold Schipper
Bildbearbeitung/Photo editing, Balázs Farkas, Gyúro, Hungary
Druck und Bindung/Printing and Binding: Westermann Druck Zwickau, GmbH, Germany

ISBN 978-3-930612-99-4

Contents, distribution and IUCN Red List data (abbreviations see p. 8)/ **Inhalt, Verbreitungsangaben und Gefährdung laut IUCN** (Abk. vgl. S. 8)

Introduced to Australia (at least New South Wales, Queensland and Victoria), French Polynesia, Guam, Hawaii (at least Kaua'i, Maui and O'ahu Islands; USA), Indonesia (New Guinea), Micronesia, New Zealand, the Northern Mariana Islands (Saipan Island), Palau, Papua New Guinea, Japan (Ogasawara Islands) and probably additional countries/Eingeführt nach Australien (mindestens New South Wales, Queensland und Victoria), Französisch-Polynesien, Guam, Hawaii (mindestens auf den Inseln Kaua'i, Maui und O'ahu; USA), Indonesien (Neuguinea), Mikronesien, Neuseeland, auf den Nördlichen Marianen (Insel Saipan), Palau, nach Papua-Neuguinea, Japan (Ogasawara-Inseln) und vermutlich in weitere Länder

EMYDINAE Rafinesque, 1815 – Holarctic Pond Turtles/Holarktische Sumpfschildkröten

Introduced species/eingeführte Arten

Actinemys marmorata (Baird & Girard, 1852) – Northern Pacific Pond Turtle/Nördliche Pazifische Sumpfschildkröte; IUCN: VU **(TERRALOG Vol. 2/Band 2)**

Introduced to Australia (New South Wales)/Eingeführt in Australien (New South Wales)

Terrapene carolina (Linnaeus, 1758) – Eastern Box Turtle/Gewöhnliche Dosenschildkröte (subspecies *T. c. triunguis* [Agassiz, 1857]/Unterart *T. c. triunguis* [Agassiz, 1857]); IUCN: VU **(TERRALOG Vol. 2/Band 2)**

Introduced to Guam/Eingeführt auf Guam

GEOEMYDIDAE Theobald, 1868 – Old World Pond Turtles/Altwelt-Sumpfschildkröten

GEOEMYDINAE Theobald, 1868 – Old World Pond Turtles/Altwelt-Sumpfschildkröten

Introduced species/eingeführte Arten

Mauremys reevesii (Gray, 1831) – Reeves' Turtle/Chinesische Dreikielschildkröte; IUCN: EN **(TERRALOG Vol. 4/Band 4)**

Introduced to Palau and possibly Guam/Eingeführt auf Palau und möglicherweise Guam

Mauremys sinensis (Gray, 1834) – Chinese Stripe-necked Turtle/Chinesische Streifenschildkröte; IUCN: EN **(TERRALOG Vol. 4/Band 4)**

Possibly introduced to Guam/Möglicherweise eingeführt auf Guam

TRIONYCHOIDEA Gray, 1825 – Softshell Turtles/Weichschildkrötenverwandtschaft

CARETTOCHELYIDAE Boulenger, 1887 – Pig-nosed Turtles/Papua-Weichschildkröten

Carettochelys Ramsay, 1886 – Pig-nosed Turtles/Papua-Weichschildkröten

Carettochelys insculpta Ramsay, 1886 – Pig-nosed Turtle/Papua-Weichschildkröte; IUCN: VU 48

Southern New Guinea (Indonesia and Papua New Guinea) and the Daly and Alligator river systems in northwestern Northern Territory (Australia)/Südliches Neuguinea (Indonesien und Papua-Neuguinea) und Flusssysteme von Daly River und den Alligator Rivers im nordwestlichen Northern Territory (Australien)

TRIONYCHIDAE Gray, 1825 – Softshell Turtles/Echte Weichschildkröten

TRIONYCHINAE Gray, 1825 – Common Softshell Turtles/Gewöhnliche Weichschildkröten

Pelochelys Gray, 1864 – Giant Softshell Turtles/Riesen-Weichschildkröten

Pelochelys bibroni (Owen, 1853) – Southern New Guinea Giant Softshell Turtle/Bibrons Riesen-Weichschildkröte; IUCN: VU 51

Southern New Guinea from the Setakwa River in Papua (Indonesia) to the Brown-Laloki River system in Papua New Guinea; anecdotal reports from Saibai Island (Australia)/Südliches Neuguinea vom Setakwa River in Papua (Indonesien) bis zum Brown-Laloki-Flusssystem in Papua-Neuguinea; Einzelberichte von der Insel Saibai (Australien)

Pelochelys signifera Webb, 2003 – Northern New Guinea Giant Softshell Turtle/Gefleckte Riesen-Weichschildkröte; IUCN: NE 51

Lowlands of northern New Guinea from the Sepik River and Ramu River drainages in Papua New Guinea westward to the Wanggar River in Papua (Indonesia)/Tiefland des nördlichen Neuguinea von den Flusssystemen von Sepik River und Ramu River in Papua-Neuguinea westwärts bis zum Wanggar River in Papua (Indonesien)

Introduced species/eingeführte Arten

Apalone spinifera (Lesueur, 1827) – Spiny Softshell Turtle/Dornrand-Weichschildkröte (various subspecies/verschiedene Unterarten); IUCN: LC **(TERRALOG Vol. 2/Band 2)**

Introduced to Hawaii (O'ahu Island; USA)/Eingeführt auf Hawaii (Insel O'ahu; USA)

Palea steindachneri (Siebenrock, 1906) – Wattle-necked Softshell Turtle/Nackendornen-Weichschildkröte; IUCN: EN **(TERRALOG Vol. 4/Band 4)**

Introduced to Hawaii (Kaua'i and probably Maui and O'ahu Islands; USA)/Eingeführt auf Hawaii (Insel Kaua'i und vermutlich Inseln Maui und O'ahu; USA)

Pelodiscus maackii (Brandt, 1857) – Northern Chinese Softshell Turtle/Amur-Weichschildkröte; IUCN: NE **(TERRALOG Vol. 4/Band 4)**

Introduced to Hawaii (O'ahu Island; USA)/Eingeführt auf Hawaii (Insel O'ahu; USA)

Pelodiscus sinensis (Wiegmann, 1835) – Chinese Softshell Turtle/Chinesische Weichschildkröte; IUCN: VU **(TERRALOG Vol. 4/Band 4)**

Introduced to Guam, the Northern Mariana Islands, Japan (Ogasawara Islands) and Hawaii (O'ahu and probably Kaua'i and Maui Islands; USA)/Eingeführt auf Guam, den Nördlichen Marianen und in Japan (Ogasawara-Inseln) sowie auf Hawaii (Insel O'ahu und vermutlich Inseln Kaua'i und Maui; USA)

IUCN Red List categories/ Gefährdungskategorien der IUCN

EX – Extinct/ausgestorben
EW – Extinct in the Wild/in freier Wildbahn ausgestorben
CR – Critically Endangered/vom Aussterben bedroht
EN – Endangered/stark gefährdet
VU – Vulnerable/gefährdet
NT – Near Threatened/Vorwarnliste
LC – Least Concern/ungefährdet
DD – Data Deficient/Daten unzureichend
NE – Not Evaluated/nicht bewertet

IUCN (2018): IUCN Red List of Threatened Species. Version 2018.1. – www.iucnredlist.org. Downloaded on 26 October 2018.
Taxa without Red List data were not listed by IUCN so far/Taxa ohne Angaben zur Gefährdungskategorie wurden von der IUCN bislang nicht aufgelistet.

Introduction

Volume 5 of the series TERRALOG deals with the chelonians of the Australasian realm, a zoogeographic region that comprises Australia, New Guinea and New Zealand, including their offshore islands, as well as the islands in the Pacific. Defining a border in the northwest of this region is fraught with some difficulties, because there exists an overlap with the Oriental realm that is known as Wallacea. With the exception of Bali, the latter groups together the Small Sunda Islands, Sulawesi, the Moluccas, and the Philippines (without Palawan); although some representatives of the chelonian fauna of the Australasian realm have dispersed as far as the Small Sunda Islands, those of Asian origin constitute the majority there so that the Wallacea is dealt with in Volume 4 of TERRALOG (East and South Asia). This explains why the present volume covers only the Indonesian provinces of Papua and Papua Barat (West Papua) as well as the Aru Islands that form part of this country's Maluku Province. As far as the Pacific islands are concerned, the Mexican Revillagigedo Islands, Cocos Island of Costa Rica, the Colombian island of Malpelo, and the Ecuadorian Galápagos Archipelago in the eastern Pacific are ignored here for zoogeographic and political reasons and incorporated in Volume 3 of TERRALOG (Central and South America) instead, even though the first three of the mentioned islands or island groups "only" host marine turtle populations anyway.

The Australian continent once formed part of the southern supercontinent Gondwana, but has not been connected to any other landmass for some 8,000 years, which was when the Torres Strait was flooded and Australia was disconnected from New Guinea. As for Gondwana itself, the supercontinent had started falling apart some 150 million years ago when Africa and Madagascar first started separating until New Zealand split off from today's Australia some 80 million years ago and, in a final separation process, Antarctica detached from it approximately 45 million years ago. Overall, the Australian landmass has thus been isolated from any other larger landmass for longer than any other continent. Terrestrial vertebrates that since have evolved could therefore not normally reach Australia, for which reason a unique fauna could evolve and persist here, and this evolutionary history also manifests in its chelonian fauna.

The chelonian fauna of the Australasian realm is by far dominated by the family of the side-necked turtles (Chelidae) that have become most speciose here, whereas families that are widely distributed in other parts of the world, namely tortoises (Testudinidae) and Old World terrapins (Geoemydidae) are absent. Present-day endemites, i.e., chelonians whose distribution is exclusive to the Australasian realm today, are limited to the family of the Papua soft-shelled turtles (Carettochelyidae). It comprises only one extant species that is distributed in the southern parts of New Guinea and northern Australia. The chelonians of New Guinea altogether exhibit very close ties to those occurring in Australia, which is a result of the long-existing land bridge. On the other hand, only one genus of true soft-shelled turtles (Trionychidae) has reached this island via the Wallacea from the Oriental realm.

Einführung

Band 5 der TERRALOG-Reihe befasst sich mit den Schildkröten der Australis, einer zoogeografischen Region, zu der Australien, Neuguinea und Neuseeland einschließlich jeweils vorgelagerter Inseln sowie die Inselwelt des Pazifik gezählt werden. Die Grenzziehung im Nordwesten bereitet etwas Schwierigkeiten, da dort ein Mischgebiet mit der orientalischen Faunenregion (Orientalis) besteht, das als Wallacea bezeichnet wird und mit Ausnahme von Bali die Kleinen Sundainseln und außerdem Sulawesi, die Molukken und die Philippinen (ohne Palawan) umfasst; zwar haben aus der Schildkrötenfauna auch Vertreter der australischen Region den „Sprung" bis auf die Kleinen Sundainseln geschafft, doch überwiegen in diesem Gebiet die Schildkröten asiatischen Ursprungs, so dass die Wallacea mit in Band 4 der TERRALOG-Reihe (Ost- und Südasien) abgehandelt wird. So werden vom indonesischen Staatsgebiet im vorliegenden Band lediglich die Provinzen Papua und Papua Barat (Westpapua) sowie die zur Provinz Maluku zählenden Aru-Inseln abgedeckt. Was die Inseln des Pazifiks betrifft, so werden die mexikanischen Revillagigedo-Inseln, die zu Costa Rica gehörende Kokos-Insel, die kolumbianische Insel Malpelo und der ecuadorianische Galápagos-Archipel im Ostpazifik aus tiergeografisch-politischen Gründen nicht in diesem Band, sondern in Band 3 der TERRALOG-Reihe (Mittel- und Südamerika) berücksichtigt, wobei die drei zuerst genannten Inseln bzw. Inselgruppen ohnehin „nur" Populationen von Meeresschildkröten aufweisen.

Der australische Kontinent war einst Teil des südlichen Superkontinents Gondwana, steht aber seit ca. 8000 Jahren mit keiner anderen Landmasse mehr in Verbindung; zu diesem Zeitpunkt wurde durch die Überflutung der Torres-Straße die Verbindung von Australien und Neuguinea getrennt, nachdem vor etwa 150 Millionen Jahren Gondwana zu zerbrechen begann (zunächst zwischen Afrika und Madagaskar) und sich vor ungefähr 45 Millionen Jahren zuletzt die Antarktis und vor ca. 80 Millionen Jahren Neuseeland vom heutigen Australien gelöst hatten. Insgesamt war der australische Erdteil somit länger als irgendein anderer Kontinent von anderen größeren Landmassen isoliert. Landwirbeltiere, die sich seitdem herausbildeten, besiedelten Australien daher im Allgemeinen nicht, weshalb sich hier eine einzigartige und relativ ursprüngliche Tierwelt entwickeln bzw. erhalten konnte, was sich auch im Bereich der Schildkröten zeigt.

Die Schildkrötenfauna der Australis gehört ganz überwiegend der Familie der Schlangenhalsschildkröten (Chelidae) an, die hier ihr Mannigfaltigkeitszentrum besitzt, während sonst weit verbreitete Familien wie die Landschildkröten (Testudinidae) und die Altwelt-Sumpfschildkröten (Geoemydidae) fehlen. Endemisch, d. h. in ihrer heutigen Verbreitung vollständig auf die Australis beschränkt, ist unter den Schildkröten lediglich die Familie der Papua-Weichschildkröten (Carettochelyidae), die nur eine noch heute existierende Art umfasst, die im Süden Neuguineas und im Norden Australiens verbreitet ist. Die Schildkröten Neuguineas weisen aufgrund der lange bestehenden Landver-

Both New Zealand and the numerous smaller Pacific islands do not host any terrestrial or fresh water chelonians today other than those populations that were introduced by man on various occasions. Their shore waters are home to marine turtles, though, that will frequent many Pacific islands for the purpose of oviposition. The Australasian realm is furthermore the only zoogeographic region in which an endemic species of marine turtle is found: the flatback turtle (*Natator depressus*). The last representatives of the terrestrial genus *Meiolania* (family Meiolaniidae), which used to exist in Australia, on Lord Howe Island, New Caledonia, Vanuatu, and presumably also Fiji, were eradicated here in prehistoric and historic times, only a few centuries after man had first arrived here; it is for example generally presumed that the species *Meiolania damelipi* that once lived on the island of Efate, which today belongs to the state of Vanuatu, fell victim to the hunters of the Lapita people only some 2,800 years before present (WHITE *et al.* 2010).

I here follow with my list of taxa at least in part the presently valid systematic list of the TURTLE TAXONOMY WORKING GROUP (2007), including its subsequent revisions by RHODIN *et al.* (2008) and the TURTLE TAXONOMY WORKING GROUP (2009, 2010, 2011, 2012, 2014, 2017), but the volume by CANN & SADLIER (2017) in the cases of various Australian chelids that have not yet been considered for the latest issue by the TURTLE TAXONOMY WORKING GROUP (2017). In those cases where the TURTLE TAXONOMY WORKING GROUP (2017) and CANN & SADLIER (2017) differ in opinion with regard to the nomenclatural availability of names introduced by Richard WELLS, Ross WELLINGTON and Raymond HOSER in various publications (see the explanations in the respective species accounts), the evaluation of the situation by the TURTLE TAXONOMY WORKING GROUP is given preference. Distribution data have for the largest part been adopted from GEORGES & THOMSON (2010), the TURTLE TAXONOMY WORKING GROUP (2017), and CANN & SADLIER (2017).

Deviating from previous volumes of TERRALOG, the listing of as yet undescribed "forms" has been done without in the present volume. However, those genera that presumably contain undescribed taxa according to the current state of knowledge are briefly discussed in the systematic section. Species that were introduced by man to areas forming part of the geographic region dealt with here are mentioned in the text as well.

Systematic arrangements are portrayed in a highly simplified manner here. Some "modern" systematic lists contain additional intermediate categories such as infraorders, genus groups, and others more that are in part based upon the inclusion of now-extinct taxonomic groups of chelonians. These categories may well be warranted, but they are ignored here for the benefit of a simplified and uncluttered systematic overview. As a consequence, the sequence of superfamilies, families, subfamilies, genera and species is given in an order that may not always correctly reflect their factual cladistic relationships.

bindung mit Australien insgesamt sehr enge Beziehungen zu den dort verbreiteten Arten auf, von der Orientalis dringt über die Wallacea lediglich eine Gattung der Echten Weichschildkröten (Trionychidae) bis zu zu dieser Insel vor.

Auf Neuseeland und den zahlreichen kleineren Inseln des Pazifiks existieren, wenn man von verschiedentlich vom Menschen eingeführten Populationen absieht, heute weder Land- noch Süßwasserschildkröten, doch sind in deren Küstengewässern Meeresschildkröten zuhause, die viele der pazifischen Inseln auch zur Eiablage aufsuchen. Die Australis ist zudem die einzige zoogeografische Region, in der eine endemische Meeresschildkrötenart beheimatet ist, die Flache Meeresschildkröte (*Natator depressus*). Die letzten Vertreter der terrestrisch lebenden Gattung *Meiolania* (Familie Meiolaniidae), die einst in Australien, auf der Lord-Howe-Insel, auf Neukaledonien, Vanuatu und vermutlich Fidschi beheimatet waren, wurden erst in vor- und frühgeschichtlicher Zeit ausgerottet, wenige Jahrhunderte nach dem ersten Kontakt mit dem Menschen; so geht man z. B. davon aus, dass die einst auf der zum heutigen Vanuatu gehörenden Insel Efate lebende *Meiolania damelipi* vor etwa 2800 Jahren den Jagdaktivitäten des Stammes der Lapita zum Opfer fiel (WHITE *et al.* 2010).

Ich folge hier mit der Auflistung der Taxa z. T. der aktuellen systematischen Liste der TURTLE TAXONOMY WORKING GROUP (2007) bzw. deren durch RHODIN *et al.* (2008) und die TURTLE TAXONOMY WORKING GROUP (2009, 2010, 2011, 2012, 2014, 2017) aktualisierten Ausgaben, im Falle verschiedener australischer Cheliden jedoch dem Band von CANN & SADLIER (2017), der in der aktuellen Arbeit der TURTLE TAXONOMY WORKING GROUP (2017) noch keine Berücksichtigung finden konnte. In jenen Fällen, in denen die TURTLE TAXONOMY WORKING GROUP (2017) und CANN & SADLIER (2017) bezüglich der nomenklatorischen Verfügbarkeit einiger Veröffentlichungen von RICHARD WELLS, ROSS WELLINGTON und RAYMOND HOSER unterschiedliche Ansichten vertreten (siehe Erläuterungen in den jeweiligen Artkapiteln), wird der Einschätzung der TURTLE TAXONOMY WORKING GROUP Vorrang eingeräumt. Die Verbreitungsangaben sind weitestgehend GEORGES & THOMSON (2010), der TURTLE TAXONOMY WORKING GROUP (2017) und CANN & SADLIER (2017) entnommen.

Im Unterschied zu Vorgängerbänden der TERRALOG-Reihe wurde im vorliegenden Band auf die Aufzählung noch nicht beschriebener „Formen" verzichtet; im systematischen Teil wird jedoch bei jenen Gattungen, die nach derzeitigem Kenntnisstand vermutlich noch unbeschriebene Taxa umfassen, auf diese Thematik eingegangen. Im Text erwähnt werden ebenso in das im vorliegenden Band behandelte Gebiet vom Menschen eingeschleppte Arten.

Die hier verwendete Systematik wurde stark vereinfacht dargestellt. In „moderneren" systematischen Listen werden verschiedentlich zusätzlich noch zahlreiche Zwischenkategorien wie Zwischenordnungen, Gattungsgruppen usw. aufgeführt, die z. T. daraus resultieren, dass ausgestorbene Schildkrötengruppen mit in die Listen aufgenommen wurden. Diese Kategorien mögen auch ihre Berechtigung haben, doch soll hier zugunsten einer vereinfachten, übersichtlichen Systematik auf sie verzichtet werden. Daher wurde bei der Reihenfolge der Überfamilien, Familien, Unterfamilien, Gattungen und Arten eine Abfolge gewählt, die nicht immer den tatsächlichen verwandtschaftlichen Verhältnissen entspricht.

Order Testudines

DUBOIS & BOUR (2010) argued that the name Testudines BATSCH, 1788 as is commonly used in the existing literature for the order of the chelonians could not be applied to this systematic level, because BATSCH (1788) had clearly intended it for use at family level. They arrived at the conclusion that no valid scientific name existed at present for the order of the chelonians. The TURTLE TAXONOMY WORKING GROUP (2010, 2011, 2012, 2014, 2017) has nevertheless continued applying the name Testudines BATSCH, 1788 to the order of the chelonians at least preliminarily, reasoning that the *International Code of Zoological Nomenclature* (INTERNATIONAL COMMISSION ON ZOOLOGICAL NOMENCLATURE 1999) in their currently valid form would not apply to this systematic level, and because the discussions revolving around the proper naming of this order and the uncertainties regarding alternative interpretations had not yet produced any consensus. The TURTLE TAXONOMY WORKING GROUP (2010, 2011, 2012, 2014, 2017) furthermore did without treating chelonians in their checklist as a separate class of vertebrated animals owing to their special positioning within the animal kingdom in the manner promoted by some other authors (e.g., COLLINS *et al.* 2010). Even though the class Reptilia may indeed be paraphyletic (a group has a common ancestor, but it does not include all groups that have evolved from this ancestor) with regard to the birds and cannot therefore be maintained in its current form, the systematic consequences for the group of the chelonians resulting from this situation have as yet remained unclear, as various authors have arrived at various results with regard to the cladistic relationships between chelonians and other groups of reptiles (e.g., LAURIN & REISZ 1995, DEBRAGA & RIEPPEL 1997, MODESTO & ANDERSON 2004, BHULLAR & BEVER 2009). It has only been recently that investigations produced the unequivocal result that chelonians represented a sister group to the Archosauria, which includes crocodiles, birds and others (CRAWFORD *et al.* 2012, LOURENÇO *et al.* 2012, ZHUO *et al.* 2013).

Ordnung Testudines

DUBOIS & BOUR (2010) argumentieren, dass die in der Literatur häufig für die Ordnung der Schildkröten verwendete Bezeichnung Testudines BATSCH, 1788 nicht auf dieser systematischen Ebene verwendet werden könne, da sie von BATSCH (1788) eindeutig auf Familienebene benutzt worden sei, und kommen zu dem Ergebnis, dass für die Ordnung der Schildkröten derzeit keine gültige wissenschaftliche Bezeichnung existiert. Die TURTLE TAXONOMY WORKING GROUP (2010, 2011, 2012, 2014, 2017) verwendet den Namen Testudines BATSCH, 1788 dagegen zumindest vorläufig weiter für die Ordnung der Schildkröten, da die *Internationalen Regeln für die Zoologische Nomenklatur* (INTERNATIONAL COMMISSION ON ZOOLOGICAL NOMENCLATURE 1999) in ihrer aktuellen Fassung keine Anwendung auf diese systematische Ebene fänden und sie die Diskussion um die korrekte Bezeichnung angesichts der Unsicherheiten und alternativen Interpretationsmöglichkeiten zu dieser Thematik noch nicht für abgeschlossen halte. Auch verzichtet die TURTLE TAXONOMY WORKING GROUP (2010, 2011, 2012, 2014, 2017) in ihrer Checkliste darauf, die Schildkröten aufgrund ihrer Sonderstellung im Tierreich als eigenständige Wirbeltierklasse zu führen, wie dies z. T. von anderen Autoren getan wird (z. B. COLLINS *et al.* 2010). Zwar ist die Klasse der Reptilien in Bezug auf die Vögel offenbar paraphyletisch (paraphyletisch = eine Gruppe hat zwar eine gemeinsame Stammform, enthält aber nicht alle Gruppen, die sich von dieser Stammform herleiten) und kann daher in ihrer bisherigen Form nicht mehr aufrechterhalten werden, doch sind die daraus resultierenden systematischen Konsequenzen für die Gruppe der Schildkröten nach wie vor unklar, da verschiedene Autoren bezüglich der Verwandtschaft der Schildkröten mit anderen Reptiliengruppen zu unterschiedlichen Ergebnissen kommen (z. B. LAURIN & REISZ 1995, DEBRAGA & RIEPPEL 1997, MODESTO & ANDERSON 2004, BHULLAR & BEVER 2009), wobei neuere Untersuchungen übereinstimmend feststellen, dass die Schildkröten eine Schwestergruppe der Archosauria darstellen, denen u. a. die Krokodile und Vögel angehören (CRAWFORD *et al.* 2012, LOURENÇO *et al.* 2012, ZHUO *et al.* 2013).

Family Cheloniidae

The positioning of the genera of the family Cheloniidae as per their degrees of relatedness has seen various approaches ever since GRAY (1825) first suggested an inadequately defined family "Carettidae" next to the Cheloniidae. Some authors have since subdivided the true marine turtles into subfamilies (e.g., DERANIYAGALA 1952, MŁYNARSKI 1976, PRITCHARD & TREBBAU 1984, SMITH & SMITH 1980) or even different families (DERANIYAGALA 1933), but others have accepted only groups of genera (tribes) (ZANGERL & TURNBULL 1955, ZANGERL 1958). These different concepts are mainly the result of the uncertain positions of the genera *Eretmochelys* and *Natator* within the family as well as the

Familie Cheloniidae

Bei der Anordnung der Gattungen der Familie Cheloniidae in Verwandtschaftsgruppen werden seit GRAY (1825), der erstmals neben den Cheloniidae eine unklar definierte Familie „Carettidae" aufführte, unterschiedliche Ansätze verfolgt. Zum Teil werden die Echten Meeresschildkröten seitdem in Unterfamilien (z. B. DERANIYAGALA 1952, MŁYNARSKI 1976, PRITCHARD & TREBBAU 1984, SMITH & SMITH 1980) oder gar Familien (DERANIYAGALA 1933), z. T. auch nur in Gattungsgruppen (Triben) (ZANGERL & TURNBULL 1955, ZANGERL 1958) unterteilt. Diese verschiedenen Ansätze basieren vor allem auf der unsicheren Position der Gattungen *Eretmochelys* und *Natator* innerhalb der Familie sowie der unresolved cladistic relationships between fossil and recent species (BOWEN *et al.* 1993, PARHAM & FASTOVSKY 1997). Using different methods, four studies of late have brought a little more clarity into this subject (NARO-MACIEL *et al.* 2008, PARHAM & PYENSON 2010, DUCHENE *et al.* 2012, GUILLON *et al.* 2012) in that they all arrived at the conclusion that the genus *Natator* is sister to the genus *Chelonia*, whereas the genus *Eretmochelys* has the same relationship to the genera *Caretta* and *Lepidochelys*. The consequences drawn from of these realisations are again different, however, ranging from the acceptance of subfamilies (DUCHENE *et al.* 2012) or tribes (NARO-MACIEL *et al.* 2008) via the recognition of a tribe for genera *Caretta* and *Lepidochelys* (Carettini) without further subdivision of the family Cheloniidae (PARHAM & FASTOVSKY 1997), to an abandonment of the idea of intrafamilial grouping (GUILLON *et al.* 2012). Pending the clarification of the cladistic relationships between fossil and recent species based on these more up-to-date research results, the TURTLE TAXONOMY WORKING GROUP (2012, 2014, 2017) has been treating the identified groupings as subfamilies of the family Cheloniidae.

unklaren Verwandtschaftsbeziehungen der fossilen zu den rezenten Arten (BOWEN *et al.* 1993, PARHAM & FASTOVSKY 1997). Vier neuere Untersuchungen bringen mit unterschiedlichen Methoden mehr Klarheit in diese Fragestellung (NARO-MACIEL *et al.* 2008, PARHAM & PYENSON 2010, DUCHENE *et al.* 2012, GUILLON *et al.* 2012). Alle kommen zu dem Ergebnis, dass die Gattung *Natator* in einem Schwestergruppenverhältnis zur Gattung *Chelonia* steht, die Gattung *Eretmochelys* dagegen in einem solchen zu den Gattungen *Caretta* und *Lepidochelys*. Die Konsequenzen, die daraus gezogen werden, sind allerdings erneut unterschiedlich und reichen von der Anerkennung von Unterfamilien (DUCHENE *et al.* 2012) bzw. Triben (NARO-MACIEL *et al.* 2008) über die Zuerkennung des Status' eines Tribus für die Gattungen *Caretta* und *Lepidochelys* (Carettini) ohne eine weitere Unterteilung der Familie Cheloniidae (PARHAM & FASTOVSKY 1997) bis hin zum Verzicht auf den Versuch einer Schaffung von innerfamiliären Gruppierungen (GUILLON *et al.* 2012). Die TURTLE TAXONOMY WORKING GROUP (2012, 2014, 2017) führt die identifizierten Gruppierungen innerhalb der Familie Cheloniidae bis zu einer weiteren Klärung der Verwandtschaftsverhältnisse von fossilen und rezenten Arten in Anerkennung der neueren Forschungserkenntnisse als Unterfamilien auf.

Genus *Caretta*:
While SHAMBLIN *et al.* (2014) were able to identify in the framework of genetic studies a total of 59 haplotypes in specimens from 42 nesting beaches, they refrained from suggesting any taxonomical changes.

Gattung *Caretta*:
SHAMBLIN *et al.* (2014) konnten im Rahmen ihrer genetischen Studien zwar an Exemplaren von 42 Niststränden insgesamt 59 Haplotypen identifizieren, zogen daraus allerdings keine Konsequenzen taxonomischer Art.

Genus *Eretmochelys*:
According to HALLERMANN (2007), Carolus LINNAEUS was officially ennobled in 1761 and subsequently called himself Carl VON LINNÉ, so that the works he published after this point of time should correctly be credited to this form of his name.
Although the results of genetic studies indicate that the populations of the hawksbill turtle in the Atlantic, Pacific and Indian Ocean are distinctly different from each other, no recent reviser has propagated the continued differentiation of the two subspecies *imbricata* and *bissa* that used to be widely practised before (OKAYAMA *et al.* 1999, BOWEN & KARL 2007, DUCHENE *et al.* 2012, GAOS *et al.* 2016).

Gattung *Eretmochelys*:
Nach HALLERMANN (2007) wurde CAROLUS LINNAEUS im Jahre 1761 offiziell geadelt und bezeichnete sich danach selbst als CARL VON LINNÉ, sodass Arbeiten, die er nach diesem Zeitpunkt veröffentlichte, unter diesem Namen zitiert werden sollten.
Während die Ergebnisse genetischer Studien zwar darauf hindeuten, dass sich die Populationen der Echten Karettschildkröte in Atlantik, Pazifik und Indischem Ozean genetisch deutlich voneinander unterscheiden, sprechen sich keine der aktuellen Bearbeiter dafür aus, die bislang meist übliche Untergliederung der Art in die beiden Unterarten *imbricata* und *bissa* aufrechtzuerhalten (OKAYAMA *et al.* 1999, BOWEN & KARL 2007, DUCHENE *et al.* 2012, GAOS *et al.* 2016).

Genus *Lepidochelys*:
FLORES-VILLELA *et al.* (2016) have demonstrated that *Lepidochelys olivacea* was in fact first described in a contribution to the periodical "*Die Quatember*" (ESCHSCHOLTZ 1829a) and not, as has been supposed thus far in the "*Zoologischer Atlas*" (ESCHSCHOLTZ 1829b).

Gattung *Lepidochelys*:
FLORES-VILLELA *et al.* (2016) zeigen auf, dass *Lepidochelys olivacea* in einem Beitrag in der Zeitschrift „*Die Quatember*" (ESCHSCHOLTZ 1829a) erstmals beschrieben wurde, und nicht wie bisher vermutet im „*Zoologischen Atlas*" (ESCHSCHOLTZ 1829b).

Genus *Chelonia*:
The study conducted by BOWEN *et al.* (1992) revealed that a recognition of the taxon "*Chelonia agassizii* BOCOURT, 1868" would render the taxon *Chelonia mydas* (LINNAEUS, 1758) paraphyletic. It is for this reason that *agassizii* is largely no longer accepted as a separate species or subspecies. Detailed

Gattung *Chelonia*:
Die Untersuchung von BOWEN *et al.* (1992) zeigt, dass die Anerkennung des Taxons „*Chelonia agassizii* BOCOURT, 1868" das Taxon *Chelonia mydas* (LINNAEUS, 1758) paraphyletisch machen würde. Aus diesem Grund wird *agassizii* überwiegend nicht länger als eigenständige Art oder Unter-

discussions of this subject can be found in PARHAM & ZUG (1996) and KARL & BOWEN (1999). In contrast, OKAMOTO & KAMEZAKI (2014) differentiated on the basis of the colour pattern of the shell and head and shell morphology between a "yellow" and a "black" form of *Chelonia mydas* that live in the western Pacific off the coast of Japan. They supposed that the "black" form represented "*Chelonia agassizii* BOCOURT, 1868" and accepted this denomination at species level. However, referring to the works cited above, the TURTLE TAXONOMY WORKING GROUP (2017) has opted for leaving *agassizii* in the synonymy of *Chelonia mydas* until more studies on the morphology and genetics of these turtles would become available. BOWEN & KARL (2007) noted genetic differences between the populations in the Atlantic and Pacific/Indian Ocean, respectively, DUCHENE *et al.* (2012) between those in the Pacific and in the Atlantic/Indian Ocean, and NARO-MACIEL *et al.* (2014) finally between those in the South and North Atlantic, but all teams of authors refrained from recognising subspecies. Likewise, SHAMBLIN *et al.* (2012) did not draw taxonomical consequences from the result of their study that nesting populations of the common green turtle from Buck Island (American Virgin Islands), the Venezuelan island of Aves, from Suriname, and those from Tortuguero (Costa Rica) could be distinguished genetically and should for this reason be treated as separate "management units" from a conservation perspective.

Family Dermochelyidae
Genus *Dermochelys*:
DUCHENE *et al.* (2012) investigated the variability of the mitochondrial genome of the leatherback turtle and noted distinct differences between the Pacific populations on the one, and those of the Atlantic and Indian Ocean on the other hand, but did not suggest corresponding taxonomic changes. DUTTON *et al.* (2013) and MOLFETTI *et al.* (2013) proved the existence of a distinct genetic structure within the Atlantic populations, but likewise refrained from taking taxonomic steps.

Family Carettochelyidae
Genus *Carettochelys*:
WELLS (2002a) described a weakly defined subspecies of *Carettochelys insculpta* under the denomination "*Carettochelys insculpta canni*" and restricted the occurrence of the "nominate subspecies" to the south of New Guinea. THOMSON & GEORGES (2009) generally regard the taxa introduced by Richard WELLS in his self-published online journal *Australian Biodiversity Record* as non-conforming with the *International Code of Zoological Nomenclature* and therefore as nomenclaturally unavailable. This also applies to the denomination "*Carettochelys insculpta canni*" that was described in said publication. While "*Carettochelys insculpta canni*" was included in the systematic lists by FRITZ & HAVAŠ (2006), the TURTLE TAXONOMY WORKING GROUP (2007), and by RHODIN *et al.* (2008) at least preliminarily, this has no longer been the case in the more recent lists by the TURTLE TAXONOMY WORKING GROUP (2009, 2010, 2011, 2012, 2014,

art angesehen. Eine ausführliche Diskussion dieser Thematik findet sich bei PARHAM & ZUG (1996) und KARL & BOWEN (1999). OKAMOTO & KAMEZAKI (2014) unterscheiden zwar anhand der Panzerfärbung und Kopf- und Panzermorphologie eine „gelbe" und eine „schwarze" Form von *Chelonia mydas* im westlichen Pazifik vor der Küste Japans und gehen davon aus, dass es sich bei der „schwarzen" Form um „*Chelonia agassizii* BOCOURT, 1868" handelt und räumen ihr Artrang ein, doch betrachtet die TURTLE TAXONOMY WORKING GROUP (2017) unter Verweis auf die oben genannten Arbeiten *agassizii* bis zum Vorliegen weiterer morphologischer und genetischer Studien weiterhin als Synonym von *Chelonia mydas*. BOWEN & KARL (2007) stellen genetische Unterschiede zwischen den Populationen in Atlantik und Pazifik/Indischem Ozean fest, DUCHENE *et al.* (2012) zwischen den Beständen im Pazifik und jenen im Atlantik/Indischen Ozean sowie NARO-MACIEL *et al.* (2014) zwischen jenen im südlichen und nördlichen Atlantik, doch verzichten alle Autorenteams auf eine Anerkennung von Unterarten. Ebenso wenig ziehen SHAMBLIN *et al.* (2012) taxonomische Konsequenzen aus dem Ergebnis ihrer Studie, dass die Nistpopulationen der Suppenschildkröte der Insel Buck (Amerikanische Jungferninseln), der venezolanischen Insel Aves, aus Suriname und Tortuguero (Costa Rica) genetisch unterscheidbar sind und artenschutztechnisch als getrennte „Management Units" behandelt werden sollten.

Familie Dermochelyidae
Gattung *Dermochelys*:
DUCHENE *et al.* (2012) untersuchen die Variabilität des mitochondrialen Genoms der Lederschildkröte und stellen deutliche Unterschiede zwischen den Populationen des Pazifiks auf der einen und des Atlantiks und Indischen Ozeans auf der anderen Seite fest, doch ziehen sie daraus keine entsprechenden taxonomischen Konsequenzen. DUTTON *et al.* (2013) und MOLFETTI *et al.* (2013) weisen eine deutliche genetische Strukturierung innerhalb der Populationen des Atlantiks nach, verzichten jedoch ebenfalls auf Maßnahmen taxonomischer Art.

Familie Carettochelyidae
Gattung *Carettochelys*:
WELLS (2002a) beschreibt eine nur schwach definierte Unterart von *Carettochelys insculpta* unter der Bezeichnung „*Carettochelys insculpta canni*" und beschränkt das Vorkommen der „Nominatform" auf den Süden Neuguineas. THOMSON & GEORGES (2009) betrachten generell die von RICHARD WELLS in der von ihm selbst herausgegebenen Online-Zeitschrift *Australian Biodiversity Record* aufgestellten Taxa als nicht den *Internationalen Regeln für Zoologische Nomenklatur* entsprechend veröffentlicht und als daher nomenklatorisch nicht verfügbar. Dies gilt auch für die in jener Zeitschrift beschriebene „*Carettochelys insculpta canni*". Während in den systematischen Listen von FRITZ & HAVAŠ (2006), der TURTLE TAXONOMY WORKING GROUP (2007) bzw. von RHODIN *et al.* (2008) „*Carettochelys insculpta canni*" zumindest vorläufig noch als valides Taxon aufgeführt wird, ist dies in den aktuelleren Listen der TURTLE TAXONOMY

2017), endorsing the assessment that this name is unavailable as per the *International Code of Zoological Nomenclature*. FRITZ & HAVAŠ (2007), GEORGES *et al.* (2008) and GEORGES & THOMSON (2010) have since also spoken out against the nomenclatural availability of taxa suggested in the *Australian Biodiversity Record*.

Family Trionychidae
Genus *Pelochelys*:
Among others, HOSER (2014a) proposed a separate subgenus named *"Ferepelochelys"* for the two New Guinean species of this genus. Like the numerous others of HOSER's "descriptions", this work was published in a journal with minuscule circulation that had been created for exactly this purpose, without the peer review system that has become a must for scientific publications worldwide, and with disregard for scientific standards and the rules of ethics. This work attitude has by now resulted in some kind of "parallel taxonomy" in Australian reptiles in particular that threatens nomenclatural stability because, unfortunately, some of the questionable descriptions by HOSER just about conform to the *International Code of Zoological Nomenclature* from a purely formal perspective, so that they had to be accepted by the scientific community at least in some instances and theoretically. Scientists, amateurs, and all larger herpetological societies have since formed a wide front in an effort to stop HOSER's practices (KAISER *et al.* 2013, KAISER 2014, KWET 2014, VETTER 2014, RHODIN *et al.* 2015). Triggered by a supplication by RHODIN *et al.* (2015), the *International Commission for Zoological Nomenclature* (ICZN) is currently looking into this matter in search of a possible ruling on how the taxa proposed by HOSER must be dealt with in general – this ruling is eagerly anticipated. None of the taxa introduced by HOSER are recognised in the present volume, but they are mentioned for completeness sake where the taxonomies of the individual genera are discussed.

Family Chelidae
Recent years saw various authors (MICHAEL & LINDERMEYER 2010, WILSON & SWAN 2013) applying the grammatically correctly derived spelling "Cheluidae" for this family. They followed STORR (1978) who had argued that it was based on the South American type genus *Chelus*. The TURTLE TAXONOMY WORKING GROUP (2017) pointed out, however, that, according to the *International Code of Zoological Nomenclature*, the oldest spelling should be maintained – even if it were "wrong" in a grammatical sense – if such spelling had been widely and continuously used, which has been the case in "Chelidae" since the twenties of the 19th century. I here follow the TURTLE TAXONOMY WORKING GROUP (2017) and retain this spelling for the present volume.
GAFFNEY (1977) surmised that a closer relationship between the genera comprising species with particularly long necks (*Chelodina* and the South American *Chelus* and *Hydromedusa*) would be likely. PRITCHARD (1984), on the other hand, attested that these three genera need not necessarily be more closely related, as the evolution of a long neck occurred in

WORKING GROUP (2009, 2010, 2011, 2012, 2014, 2017) nicht mehr der Fall, da nun auch hier der Name als nach den *Internationalen Regeln für Zoologische Nomenklatur* nicht verfügbar behandelt wird. Auch FRITZ & HAVAŠ (2007), GEORGES *et al.* (2008) und GEORGES & THOMSON (2010) sprechen sich gegen die nomenklatorische Verfügbarkeit der im *Australian Biodiversity Record* aufgestellten Taxa aus.

Familie Trionychidae
Gattung *Pelochelys*:
HOSER (2014a) stellt u. a. für die beiden auf Neuguinea verbreiteten Arten dieser Gattung eine eigenständige Untergattung mit dem Namen *„Ferepelochelys"* auf. Ebenso wie die zahlreichen anderen „Neubeschreibungen" HOSERS erschien jedoch auch diese Arbeit in einem eigens zu diesem Zweck produzierten Magazin in winziger Auflage, ohne das weltweit in der Wissenschaft übliche Peer-Review-System und unter Verstoß gegen wissenschaftliche Standards und die Regeln der Ethik. Diese Arbeitsweise hat insbesondere bei den australischen Reptilien mittlerweile zu einer Art „Paralleltaxonomie" geführt, die die nomenklatorische Stabilität gefährdet, da leider manche der fragwürdigen Beschreibungen von HOSER rein formal gerade noch den *Internationalen Regeln für Zoologische Nomenklatur* entsprechen, sodass sie zumindest teilweise theoretisch in der Wissenschaftsgemeinde anerkannt werden müssten. Um HOSERS Treiben Einhalt zu gebieten, hat sich eine breite Front aus Wissenschaftlern, Laien und allen großen herpetologischen Gesellschaften gebildet (KAISER *et al.* 2013, KAISER 2014, KWET 2014, VETTER 2014, RHODIN *et al.* 2015). So beschäftigt sich die *Internationalen Kommission für Zoologische Nomenklatur* (ICZN) derzeit auf Antrag von RHODIN *et al.* (2015) mit der Frage, wie die von HOSER publizierten Taxa insgesamt zu behandeln sind – eine Entscheidung, die mit Spannung erwartet wird. Im vorliegenden Band werden die von HOSER beschriebenen Taxa insgesamt nicht anerkannt, der Vollständigkeit halber jedoch im Rahmen der taxonomischen Hinweise zu den jeweiligen Gattungen erwähnt.

Familie Chelidae
In den vergangenen Jahren verwendeten verschiedene Autoren (MICHAEL & LINDERMEYER 2010, WILSON & SWAN 2013) für diese Familie die bereits von STORR (1978) eingeführte Schreibweise „Cheluidae", grammatikalisch korrekt abgeleitet von der südamerikanischen Typusgattung *Chelus*. Die TURTLE TAXONOMY WORKING GROUP (2017) weist jedoch darauf hin, dass gemäß den *Internationalen Regeln für Zoologische Nomenklatur* bei Familiennamen die älteste Schreibweise – auch wenn sie grammatikalisch „falsch" sein sollte – dann beizubehalten ist, wenn sie ganz überwiegend verwendet wird, was bei „Chelidae" seit den 20er-Jahren des 19. Jahrhunderts der Fall ist. Ebenso wie die TURTLE TAXONOMY WORKING GROUP (2017) behalte ich daher diese Schreibweise im vorliegenden Band bei.
GAFFNEY (1977) hält eine nähere Verwandtschaft der Gattungen mit auffällig langem Hals (*Chelodina* und die südamerikanischen *Chelus* und *Hydromedusa*) für nahe liegend. PRITCHARD (1984) stellt dagegen fest, dass diese drei Gattungen nicht notwendigerweise eine nähere Verwandt-

different manners. Genetic analyses eventually demonstrated that the Australasian Chelidae formed one cladistic group, as did the South American genera with species with shorter necks (including *Chelus*), and that the South American genus *Hydromedusa* represented a third grouping (SEDDON *et al.* 1997, GEORGES *et al.* 1998); GEORGES *et al.* (1998) suggested placing all three groups in separate subfamilies and selected the subfamily denomination Chelodininae for the Australasian clade. This viewpoint based upon genetic studies is supported by the results of morphological examinations that revealed differences between the "long-necked" genera with regard to the build of the cranium, shell and cervical vertebrae and thus indicated that the superficially similar appearance of side-necked turtles with long necks was a result of independent evolutionary processes that have shaped morphological traits in a similar manner due to similar ecological pressures rather than a sign of closer relationships (PRITCHARD 1984, THOMSON 2000). These results found corroboration in several studies in recent years, like, e.g., GUILLON *et al.* (2012) who provided a summary. Accordingly, the TURTLE TAXONOMY WORKING GROUP (2012, 2014) has also been using the three mentioned subfamilies in their checklists. In the framework of their genetic analyses, ZHANG *et al.* (2017) arrived at the conclusion that the Australian genus *Pseudemydura* was sister to all other Australasian genera of Chelidae and deserving of the rank of a separate subfamily, which had already been presumed by GAFFNEY (1977) on the basis of a single morphological character state and suggested by the results of various subsequent studies with different approaches (SEDDON *et al.* 1997, GEORGES *et al.* 1998, GUILLON *et al.* 2012, PEREIRA *et al.* 2017). This concept has therefore also been adopted for the latest checklist by the TURTLE TAXONOMY WORKING GROUP (2017).

schaft aufweisen müssen, da die Entwicklung des langen Halses auf jeweils unterschiedliche Art und Weise erfolgte. Genetische Untersuchungen zeigen so auch, dass die australasiatischen Chelidae insgesamt ebenso eine Verwandtschaftsgruppe bilden wie die südamerikanischen Gattungen mit kürzerem Hals (einschließlich *Chelus*), und dass die südamerikanische Gattung *Hydromedusa* eine dritte Gruppierung darstellt (SEDDON *et al.* 1997, GEORGES *et al.* 1998); GEORGES *et al.* (1998) schlagen für alle drei Gruppen den Rang einer Unterfamilie vor und wählen für die australasiatische Verwandtschaftsgruppe den Unterfamiliennamen Chelodininae. Diese auf genetischen Studien basierenden Erkenntnisse werden durch die Ergebnisse morphologischer Studien gestützt, die zwischen den „langhalsigen" Gattungen deutliche Unterschiede im Bau von Schädel, Panzer und Halswirbeln feststellen und darauf hinweisen, dass die äußerliche Ähnlichkeit der Schlangenhalsschildkröten mit langem Hals eher auf eine voneinander unabhängige, aber ähnliche Entwicklung von Körpermerkmalen aufgrund ähnlicher Lebensbedingungen zurückzuführen ist als auf eine engere Verwandtschaft (PRITCHARD 1984, THOMSON 2000). Bestätigt werden konnten diese Ergebnisse auch durch mehrere Untersuchungen aus den letzten Jahren, so z. B. zusammenfassend auch durch GUILLON *et al.* (2012). Dementsprechend führt auch die TURTLE TAXONOMY WORKING GROUP (2012, 2014) die genannten drei Unterfamilien in ihrer aktuellen Checkliste auf. ZHANG *et al.* (2017) kommen im Rahmen ihrer genetischen Untersuchung darüber hinaus zu dem Ergebnis, dass die australische Gattung *Pseudemydura* in einem Schwestergruppenverhältnis zu allen anderen australasiatischen Gattungen der Chelidae steht und den Rang einer eigenständigen Unterfamilie verdient, was schon GAFFNEY (1977) auf der Basis eines einzelnen morphologischen Merkmals vermutet und auch die Ergebnisse späterer Autoren immer wieder mit unterschiedlichen Ansätzen andeuten (SEDDON *et al.* 1997, GEORGES *et al.* 1998, GUILLON *et al.* 2012, PEREIRA *et al.* 2017). Dieses Ergebnis spiegelt sich auch in der aktuellen Liste der TURTLE TAXONOMY WORKING GROUP (2017) wider.

Subfamily Chelodininae:
The oldest available denomination for the subfamily of the Australian side-necked turtles is Hydraspidina BONAPARTE, 1836 (in the amended spelling applied to a subfamily: Hydraspidinae BONAPARTE, 1836), which is based on the type genus, *Hydraspis* BELL, 1828, a junior synonym of the genus *Chelodina* FITZINGER, 1826. BONAPARTE's concept of *Hydraspis* significantly differed from that of BELL's insofar as he did not place the type species of *Hydraspis*, "*Hydraspis longicollis*" (i.e., today's *Chelodina* [*Chelodina*] *longicollis*), in the genus *Hydraspis*, but, correctly, in the genus *Chelodina* that had been introduced ten year prior and left exclusively South American species in *Hydraspis*. It is for this reason that applying the subfamily denomination Hydraspidinae to the Australian representatives of the Chelidae would unnecessarily cause confusion. Because no strict priority rules from the *International Commission for Zoological Nomenclature* exist for taxonomic denominations at family (and subfamily) level, but only at genus and species levels, the TURTLE

Unterfamilie Chelodininae:
Die älteste für die Unterfamilie der Australischen Schlangenhalsschildkröten verfügbare Bezeichnung ist Hydraspidina BONAPARTE, 1836 (in der für eine Unterfamilie üblichen abgewandelten Schreibweise: Hydraspidinae BONAPARTE, 1836), die auf der Typusgattung *Hydraspis* BELL, 1828 basiert, einem jüngeren Synonym der Gattung *Chelodina* FITZINGER, 1826. BONAPARTES Konzept von *Hydraspis* weicht jedoch insofern erheblich von jenem von BELL ab, als er die Typusart von *Hydraspis*, „*Hydraspis longicollis*" (die heutige *Chelodina* [*Chelodina*] *longicollis*), eben nicht zu *Hydraspis*, sondern korrekterweise zur zehn Jahre zuvor beschriebenen Gattung *Chelodina* stellt und in *Hydraspis* ausschließlich südamerikanische Arten belässt. Aus diesem Grund würde die Verwendung des Unterfamiliennamens Hydraspidinae für die australischen Vertreter der Chelidae zu unnötiger Verwirrung führen. Da zudem für taxonomische Bezeichnungen auf Familien- (und Unterfamilien-)ebene keine strengen Prioritätsregeln seitens der *Internationalen*

Taxonomy Working Group (2012, 2014, 2017) has been promoting the use of the more unambiguous name Chelodininae Baur, 1893 for the side-necked turtles of Australia, New Guinea, and adjacent islands (with the exception of the genus *Pseudemydura*; Turtle Taxonomy Working Group 2017, Zhang *et al.* 2017)

Genus *Chelodina*:
Wells & Wellington (1985) redistributed the species of side-necked turtles that were known at the time and consolidated in the genus *Chelodina* across altogether three genera. They left in the genus *Chelodina* only the rather narrow-headed species *longicollis* and *novaeguineae* and placed the more broad-headed species *oblonga* (= today *colliei*), *expansa*, *rugosa* (= today *oblonga*), and *siebenrocki* (= today included in *oblonga*) in their newly proposed genus *Macrochelodina*, and the species *steindachneri* in the likewise newly proposed genus *Hesperochelodina*. Iverson *et al.* (2001) refuted the validity of a separate genus *Hesperochelodina* and returned the species *steindachneri* to *Chelodina*, but confirmed the validity of the genus *Macrochelodina*. Georges *et al.* (2002) retained *Chelodina* as a "collective genus", but identified three phylogenetic clades within this genus that they would accept at the rank of subgenera without giving them scientific names. Fritz & Havaš (2006, 2007) again listed two of these cladistic groups (*Chelodina*, *Macrochelodina*) at genus level. The third clade identified by Georges *et al.* (2002) contains only one species from the Australian southwest, i.e., *Chelodina colliei*, for which McCord & Joseph-Ouni (2007b) introduced the genus name *Macrodiremys*. Georges & Thomson (2010) summarized the history and results of taxonomic and systematic studies on the side-necked turtles with long necks of Australia, New Guinea, and adjacent islands and arrived at the conclusion that the cladistic relationships within this group would best be reflected if all species were placed in one genus with altogether three subgenera: *Chelodina*, *Macrochelodina* and *Macrodiremys*. This concept has also been adopted for the latest checklists by the Turtle Taxonomy Working Group (2010, 2011, 2012, 2014, 2017), whereas previous publications by this workgroup (Turtle Taxonomy Working Group 2007, 2009, Rhodin *et al.* 2008) circumvented an assignment of the various species only to *Chelodina* on the one, or *Chelodina*/*Macrochelodina*/*Macrodiremys* on the other hand.
The species today known under the denomination *Chelodina* (*Chelodina*) *canni* McCord & Thomson, 2002 had previously been described under the name "*Chelodina rankini*" by Wells & Wellington (1985), which, according to Iverson *et al.* (2001), did not conform to the *International Code of Zoological Nomenclature* so that "*Chelodina rankini* Wells & Wellington, 1985" constituted a so-called *nomen nudum* without any nomenclatural relevance. Wells (2007a) opposed this perception and redescribed "*rankini*", but the properly introduced species name *canni* McCord & Thomson, 2002 now had temporal priority from a nomenclatural perspective. Moreover, the second description of *rankini* was published in the controversial online journal *Australian*

Kommission für Zoologische Nomenklatur existieren, wie dies auf Gattungs- und Artebene der Fall ist, befürwortet die Turtle Taxonomy Working Group (2012, 2014, 2017) die Verwendung des eindeutigeren Namens Chelodininae Baur, 1893 für die Schlangenhalsschildkröten Australiens, Neuguineas und der benachbarten Inseln (mit Ausnahme der Gattung *Pseudemydura*; Turtle Taxonomy Working Group 2017, Zhang *et al.* 2017)

Gattung *Chelodina*:
Wells & Wellington (1985) verteilen die bis zu diesem Zeitpunkt bekannten Arten der in der Gattung *Chelodina* zusammengefassten Schlangenhalsschildkröten auf insgesamt drei Gattungen. Sie belassen lediglich die die eher schmalköpfigen Arten *longicollis* und *novaeguineae* in der Gattung *Chelodina* und stellen die eher breitköpfigen Arten *oblonga* (= heute *colliei*), *expansa*, *rugosa* (= heute *oblonga*) und *siebenrocki* (= heute zu *oblonga* gerechnet) in die von ihnen gleichzeitig neu aufgestellte Gattung *Macrochelodina*, die Art *steindachneri* in die ebenfalls neu beschriebene Gattung *Hesperochelodina*. Iverson *et al.* (2001) widerlegen die Berechtigung einer eigenständigen Gattung *Hesperochelodina* und belassen die Art *steindachneri* in *Chelodina*, bestätigen aber die Berechtigung einer Gattung *Macrochelodina*. Georges *et al.* (2002) halten zwar an einer „Sammelgattung" *Chelodina* fest, identifizieren jedoch innerhalb dieser Gattung drei phylogenetische Verwandtschaftsgruppen, denen sie Untergattungsrang zuerkennen, ohne diese allerdings mit einer wissenschaftlichen Bezeichnung zu belegen. Fritz & Havaš (2006, 2007) führen wiederum zwei dieser Verwandtschaftsgruppen (*Chelodina*, *Macrochelodina*) auf Gattungsebene auf. Die dritte von Georges *et al.* (2002) identifizierte Verwandtschaftsgruppe besteht lediglich aus einer Art, der südwestaustralischen *Chelodina colliei*. Für sie wurde von McCord & Joseph-Ouni (2007b) die Gattungsbezeichnung *Macrodiremys* aufgestellt. Georges & Thomson (2010) fassen die Geschichte und Erkenntnisse der Taxonomie und Systematik der langhalsigen Schlangenhalsschildkröten Australiens, Neuguineas und der benachbarten Inseln zusammen und kommen zu dem Ergebnis, dass die verwandtschaftlichen Verhältnisse innerhalb dieser Gruppe am besten dargestellt werden, indem alle Arten in eine Gattung mit insgesamt drei Untergattungen gestellt werden: *Chelodina*, *Macrochelodina* und *Macrodiremys*. Dieses Konzept wird auch in den aktuellen Checklisten der Turtle Taxonomy Working Group (2010, 2011, 2012, 2014, 2017) übernommen, während in den vorangegangenen Publikationen der Arbeitsgruppe (Turtle Taxonomy Working Group 2007, 2009, Rhodin *et al.* 2008) eine Zuordnung der verschiedenen Arten nur zu *Chelodina* auf der einen oder *Chelodina*/*Macrochelodina*/*Macrodiremys* auf der anderen Seite offengelassen wurde.
Die heute unter dem Namen *Chelodina* (*Chelodina*) *canni* McCord & Thomson, 2002 bekannte Art wurde bereits von Wells & Wellington (1985) unter der Bezeichnung „*Chelodina rankini*" beschrieben, was allerdings nach Iverson *et al.* (2001) nicht den *Internationalen Regeln für Zoologische Nomenklatur* entsprechend erfolgte, so dass es sich bei „*Chelodina rankini* Wells & Wellington, 1985" um ein sogenanntes

Biodiversity Record and is therefore considered unavailable for nomenclatural purposes (see above) (Fritz & Havaš 2007, Thomson & Georges 2009, Georges & Thomson 2010, Turtle Taxonomy Working Group 2010, 2011, 2012, 2014, 2017). This name would thus be unavailable even if the populations to the east of the Great Dividing Range (from where the type specimen of *rankini* originated, whereas that of *canni* was collected from the area to the west) were to turn out to be classifiable as a separate subspecies, which Cann & Sadlier (2017) thought of as justifiable, but Georges & Thomson (2010) found inappropriate.

Georges & Thomson (2010) considered *Chelodina (Chelodina) gunaleni* as identical to, and thus a junior synonym of, *Chelodina (Chelodina) novaeguineae*, but failed to support their assessment with corresponding data, for which reason the Turtle Taxonomy Working Group (2010, 2011, 2012, 2014, 2017) has been retaining this species until further study results become available.

The populations of *Chelodina (Chelodina) longicollis* to the east and west of the Great Dividing Range may possibly be distinguishable at subspecies level (Chessman 1978, Beck 1991, Hodges *et al.* 2015, Cann & Sadlier 2017).

As far as *Chelodina (Chelodina) novaeguineae* is concerned, some experts (Rhodin & Genorupa 2000, McCord & Joseph-Ouni 2007a) presumed that this denomination could possibly refer to several species, but studies to this effect are wanting as yet. Illustrations of three such potentially distinct forms can be found in Cann & Sadlier (2017).

In *Chelodina (Macrochelodina) expansa*, the populations of the Murray-Darling river system in the inland and the Mary River in southeastern Queensland were found to be genetically differentiated from those to the north of the Mary River, but not enough so to warrant distinguishing them at subspecies or even species level (Georges & Thomson 2010, Hodges *et al.* 2014). Hoser (2014b) assigned the populations from waters between the Mary River in the south and the Fitzroy River in the north (including Fraser Island) to his new "*Chelodina duboisi*" and those from southeastern Queensland south of the Conondale Mountains to his equally new subspecies "*Chelodina expansa brisbaneensis*". According to Hoser's "concept", the "nominate subspecies" of *Chelodina expansa* would be native to the Murray-Darling river system. *Chelodina expansa* and "*Chelodina duboisi*" were placed in the likewise newly erected subgenus "*Supremechelys*" (Hoser 2014b). All taxa proposed by Hoser (2014b) are considered invalid in the present volume for the reasons detailed above in the chapter on the genus *Pelochelys*.

Chelodina (Macrochelodina) kuchlingi was originally described on the basis of a singular specimen of unknown provenance that had previously been kept in captivity for many years; free-ranging populations of this form have not become known as yet, but this taxon has nevertheless often been treated as valid in various publications (e.g., Fritz & Havaš 2006, 2007, Turtle Taxonomy Working Group 2007, 2009, Rhodin *et al.* 2008). This was opposed by Georges & Thomson (2006, 2010), the Turtle Taxonomy Working Group (2010, 2011, 2012), Kennett *et al.* (2014), and Ellis & Georges (2015), who relegated *kuchlingi* to the synonymy of "*Chelodina rugosa*", which is today known as *Chelodina*

nomen nudum ohne nomenklatorische Relevanz handelt. Wells (2007a) widerspricht dieser Auffassung und beschreibt *rankini* erneut, doch hat der korrekt zustande gekommene Artname *canni* McCord & Thomson, 2002 aus nomenklatorischer Sicht zeitliche Priorität, zudem wurde die zweite Beschreibung von *rankini* in der bereits erwähnten umstrittenen Onlinezeitschrift *Australian Biodiversity Record* veröffentlicht und gilt deshalb als zu nomenklatorischen Zwecken nicht verfügbar (s. o.) (Fritz & Havaš 2007, Thomson & Georges 2009, Georges & Thomson 2010, Turtle Taxonomy Working Group 2010, 2011, 2012, 2014, 2017). Daher wäre der Name selbst dann nicht verwendbar, wenn sich die Populationen östlich der Great Dividing Range (von dort stammt das Typusexemplar von *rankini*, jenes von *canni* aus dem Bereich westlich davon) als Angehörige einer eigenständigen Unterart erweisen würden, was Cann & Sadlier (2017) für vertretbar halten, von Georges & Thomson (2010) allerdings als nicht zutreffend erachtet wird.

Georges & Thomson (2010) betrachten *Chelodina (Chelodina) gunaleni* als identisch mit und somit als jüngeres Synonym von *Chelodina (Chelodina) novaeguineae*, doch belegen sie diese Auffassung nicht mit entsprechenden Daten, weshalb die Turtle Taxonomy Working Group (2010, 2011, 2012, 2014, 2017) die Art bis zum Vorliegen weiterer Ergebnisse weiterhin aufführt.

Die Populationen von *Chelodina (Chelodina) longicollis* östlich und westlich der Great Dividing Range sind möglicherweise auf Unterartenebene zu unterscheiden (Chessman 1978, Beck 1991, Hodges *et al.* 2015, Cann & Sadlier 2017).

Im Falle von *Chelodina (Chelodina) novaeguineae* wird von manchen Fachleuten (Rhodin & Genorupa 2000, McCord & Joseph-Ouni 2007a) vermutet, dass sich unter dieser Bezeichnung mehrere Arten verbergen, doch fehlen dazu noch weitere Untersuchungen. Aufnahmen von drei solcher potenziell eigenständiger Taxa sind bei Cann & Sadlier (2017) zu finden.

Bei *Chelodina (Macrochelodina) expansa* konnten auf genetischer Grundlage die Populationen des Murray-Darling-Flusssystems im Landesinneren und des Mary River im Südosten Queenslands von jenen nördlich des Mary River unterschieden werden, was aber derzeit nicht als ausreichend für die Unterscheidung von Unterarten oder gar Arten angesehen wird (Georges & Thomson 2010, Hodges *et al.* 2014). Hoser (2014b) stellt die Populationen aus den Gewässern zwischen dem Mary River im Süden und dem Fitzroy River im Norden (einschließlich der Insel Fraser) zu der von ihm neu beschriebenen Art „*Chelodina duboisi*" und jene aus dem Südosten Queenslands südlich des Conondale-Berglandes zu der gleichzeitig beschriebenen Unterart „*Chelodina expansa brisbaneensis*". Die „Nominatform" von *Chelodina expansa* besiedelt nach dem „Konzept" Hosers das Murray-Darling-Flusssystem. *Chelodina expansa* und „*Chelodina duboisi*" werden in die ebenfalls neu von Hoser (2014b) aufgestellte Untergattung „*Supremechelys*" platziert. Aus den oben im Abschnitt zur Gattung *Pelochelys* erläuterten Gründen werden im vorliegenden Band die von Hoser (2014b) aufgestellten Taxa insgesamt nicht anerkannt.

Chelodina (Macrochelodina) kuchlingi wurde lediglich auf der

(*Macrochelodina*) *oblonga*. This evaluation has again been objected to by Gerald KUCHLING who has been arguing that *Chelodina* (*Macrochelodina*) *kuchlingi* is indeed a separate species with a relatively small distribution range in northeastern Western Australia (including the Ord River system) that might possibly need to be classified as endangered (TURTLE TAXONOMY WORKING GROUP 2014, 2017). The TURTLE TAXONOMY WORKING GROUP (2014, 2017) as well as CANN & SADLIER (2017) have preliminarily been retaining this form as a valid species until further research results will become available.

THOMSON (2000) was the first to demonstrate that the type specimen of *Chelodina* (*Macrochelodina*) *oblonga* GRAY, 1841, did not, as had long been presumed, originate from southwestern, but from northern Australia and therefore represented the species that had for decades been known under the denomination *"Chelodina rugosa* OGILBY, 1890". In an effort to minimize confusion, the same author (THOMSON 2006, 2007) formally appealed to the *International Commission for Zoological Nomenclature* to validate the younger name *"rugosa"* as the species name for the northern Australian side-necked turtle and so give it preference over the older denomination *"oblonga"*. This suggestion found the support of most scientific advisors who were invited to comment on the application (for a discussion of this subject, see, e.g., RHODIN *et al.* 2008). The INTERNATIONAL COMMISSION ON ZOOLOGICAL NOMENCLATURE (2013) decided to not follow this proposal, however. The consequences of this decision are that the northern Australian side-necked turtle that has been referred to in the literature under the specific epithet *"rugosa"* for decades now bears the name *"oblonga"*, under which name a multitude of publications exist that refer to the species of side-necked turtle from southwestern Australia that in turn is now known under the denomination *Chelodina* (*Macrodiremys*) *colliei* (KENNETT *et al.* 2014, TURTLE TAXONOMY WORKING GROUP 2014, 2017). Although ALACS (2008) noted genetic differences between the populations of *Chelodina* (*Macrochelodina*) *oblonga* in the Northern Territory and Western Australia on the one, and those from New Guinea, from the Cape York Peninsula, and the river systems on the Gulf of Carpentaria on the other hand, these were deemed insufficient for differentiating subspecies or even species at least at this stage. As for me, I here follow CANN & SADLIER (2017) who considered the existing genetic and morphological differences adequate for subdividing these populations at subspecies level (*oblonga oblonga* in Northern Territory and Western Australia, *oblonga rugosa* on New Guinea, the Cape York Peninsula, and on the Gulf of Carpentaria). The status of the northern Australian representatives of the subgenus *Macrochelodina* is all but resolved in other aspects either. THOMSON *et al.* (2000) noted distinct morphological differences between the populations native to the Kimberley region (today *oblonga walloyarrina*) and those from Arnhem Land (today *oblonga burrungandjii*), whereas the genetic analyses conducted by GEORGES *et al.* (2002) did not reveal them to be differentiated for which reason the name *walloyarrina* was placed in the synonymy of *burrungandjii* by GEORGES & THOMSON (2010) and ELLIS & GEORGES (2015). The TURTLE TAXONOMY WORKING GROUP

Basis eines einzelnen Exemplares unbekannter Herkunft aus langjähriger Gefangenschaftshaltung beschrieben; freilebende Populationen dieser Form sind bisher nicht bekannt, dennoch wurde sie in der Literatur häufig als valide geführt (z. B. FRITZ & HAVAŠ 2006, 2007, TURTLE TAXONOMY WORKING GROUP 2007, 2009, RHODIN *et al.* 2008). Von GEORGES & THOMSON (2006, 2010), der TURTLE TAXONOMY WORKING GROUP (2010, 2011, 2012), KENNETT *et al.* (2014) und ELLIS & GEORGES (2015) wird dies dagegen angezweifelt und *kuchlingi* in die Synonymie von *„Chelodina rugosa"*, der heutigen *Chelodina* (*Macrochelodina*) *oblonga*, gestellt. Letzterem wird wiederum von GERALD KUCHLING widersprochen, nach dessen Auffassung es sich bei *Chelodina* (*Macrochelodina*) *kuchlingi* offenbar um eine eigenständige Art mit einem relativ kleinen Verbreitungsgebiet im Nordosten Western Australias (einschließlich des Flusssystems des Ord River) handelt, die möglicherweise bereits als gefährdet einzustufen ist (TURTLE TAXONOMY WORKING GROUP 2014, 2017). Bis zum Vorliegen weiterer Untersuchungsergebnisse führt die TURTLE TAXONOMY WORKING GROUP (2014, 2017) diese Form ebenso wie CANN & SADLIER (2017) daher zumindest vorläufig als valide Art auf.

Bereits THOMSON (2000) konnte nachweisen, dass das Typusexemplar (Holotypus) von *Chelodina* (*Macrochelodina*) *oblonga* GRAY, 1841 nicht wie lange Zeit vermutet aus dem Südwesten, sondern aus dem Norden Australiens stammt und somit der Art zuzuordnen ist, die jahrzehntelang unter der Bezeichnung *„Chelodina rugosa* OGILBY 1890" bekannt war. Um Verwechslungen zu vermeiden, stellte der gleiche Autor (THOMSON 2006, 2007) bei der *Internationalen Kommission für Zoologische Nomenklatur* den Antrag, dem jüngeren Namen *„rugosa"* den Vorrang als Artbezeichnung für die Nordaustralische Schlangenhalsschildkröte gegenüber der älteren Bezeichnung *„oblonga"* einzuräumen, was von den meisten wissenschaftlichen Kommentatoren dieses Antrags unterstützt wurde (zur Diskussion dieses Themas siehe z. B. RHODIN *et al.* 2008). Die INTERNATIONAL COMMISSION ON ZOOLOGICAL NOMENCLATURE (2013) entschied sich jedoch dagegen, diesem Vorschlag zu folgen. Dies hat wiederum zur Folge, dass die Nordaustralische Schlangenhalsschildkröte, die über Jahrzehnte in der Literatur unter dem Artnamen *„rugosa"* bekannt war, nun den Namen *„oblonga"* tragen muss, zu dem zahlreiche Veröffentlichungen über die Schlangenhalsschildkrötenart aus dem Südwesten Australiens vorliegen, die nun unter der Bezeichnung *Chelodina* (*Macrodiremys*) *colliei* geführt wird (KENNETT *et al.* 2014, TURTLE TAXONOMY WORKING GROUP 2014, 2017). ALACS (2008) stellt zwar genetische Unterschiede zwischen den Populationen von *Chelodina* (*Macrochelodina*) *oblonga* im Northern Territory und Western Australia auf der einen und jenen von Neuguinea, von der Kap-York-Halbinsel und aus Flusssystemen am Golf von Carpentaria auf der anderen Seite fest, hält diese aber für eine Unterscheidung von Unterarten oder gar Arten zumindest derzeit für nicht ausreichend; ich folge hier jedoch CANN & SADLIER (2017), die eine Unterteilung in Unterarten aufgrund genetischer und morphologischer Unterschiede für gerechtfertigt erachten (*oblonga oblonga* im Northern Territory und in Western Australia, *oblonga rugosa* auf Neuguinea, der Kap-York-Halbinsel und am Golf von Carpentaria). Auch in anderer Hinsicht ist der Status der nordaustralischen Vertreter der Untergattung *Macrochelodina* alles andere als geklärt. So konnten THOMSON *et*

(2010, 2011, 2012, 2014, 2017), on the other hand, has been retaining *walloyarrina* as a valid taxon in their checklists at least preliminarily and until more data will become available. However, they have been doing so for this taxon as well as for *burrungandjii* at species level, as had been proposed in their original descriptions, and not as subspecies as was suggested by CANN & SADLIER (2017). The classification of both *walloyarrina* and *burrungandjii* as subspecies of *Chelodina* (*Macrochelodina*) *oblonga* in the present volume follows the concept by CANN & SADLIER (2017) until the cladistic relationships of these taxa will be comprehensively clarified. The populations of *Chelodina* (*Macrochelodina*) *oblonga rugosa* from New Guinea have occasionally been referred to in the literature as a species in its own right under the denomination "*Chelodina siebenrocki* WERNER, 1901" (e.g., RHODIN & MITTERMEIER 1976, RHODIN & GENORUPA 2000), but the genetic analyses conducted by GEORGES *et al.* (2002) demonstrated only minor differentiation and this name is now considered merely a synonym of *oblonga* (FRITZ & HAVAŠ 2006, 2007, TURTLE TAXONOMY WORKING GROUP 2007, 2009, 2010, 2011, 2012, 2014, 2017, RHODIN *et al.* 2008, GEORGES & THOMSON 2010).

While it may be questionable in some of the cases briefly discussed above whether the mentioned taxa indeed warrant recognition as separate species (or subspecies), no such doubts exist for the forms of side-necked turtles living in the Australian southwest. All but uncontroversial was over the past years which scientific name must correctly be assigned to this taxon, however, as was outlined in the discussion under *Chelodina* (*Macrochelodina*) *oblonga* above. In a quest to conserve the species name *oblonga* that had for so many years been applied to this southwestern Australian form even though it was actually based upon a specimen of the northern Australian side-necked turtle, which was also supported by SAVAGE (2007), McCORD & JOSEPH-OUNI (2007b) made the type specimen that THOMSON (2000) had designated to serve as the lectotype of *Chelodina colliei* (which was thought of as the next oldest name for this taxon at that time) the type specimen (neotype) of *Chelodina oblonga* as well, for which they proposed the genus name *Macrodiremys* in the same publication. In this manner, "*Chelodina oblonga*" with its new, now geographically "fitting" type specimen could have been "saved" for the side-necked turtles of southwestern Australia. The designation of the neotype by McCORD & JOSEPH-OUNI (2007b) was invalid, however, as such action could have been executed solely by the *International Commission for Zoological Nomenclature* due to the fact that the original type specimen (holotype) of *Chelodina oblonga* still exists and had merely been misapplied for many years due to confusion regarding its geographic origin (KUCHLING 2010). As has been explained already above, the INTERNATIONAL COMMISSION ON ZOOLOGICAL NOMENCLATURE (2013) eventually decided to strictly follow the rules of priority and fix the epithet *oblonga* for the northern Australian species, for which reason the southwestern Australian taxon now needs to be referred to by the oldest available name, *colliei*. At this stage, matters became really complicated at the very latest. When McCORD & JOSEPH-OUNI (2007b) associated the species they named "*Macrodiremys oblonga*" (they of course

al. (2000) zwischen den Populationen der Kimberley-Region (heute *oblonga walloyarrina*) und von Arnhemland (heute *oblonga burrungandjii*) deutliche morphologische Unterschiede feststellen, während GEORGES *et al.* (2002) im Rahmen ihrer genetischen Untersuchungen keine Differenzierung gelang, weshalb *walloyarrina* von GEORGES & THOMSON (2010) und ELLIS & GEORGES (2015) in die Synonymie von *burrungandjii* gestellt wird. Die TURTLE TAXONOMY WORKING GROUP (2010, 2011, 2012, 2014, 2017) führt *walloyarrina* dagegen zumindest vorläufig bis zum Vorliegen weiterer Daten in ihrer Checkliste als eigenständiges Taxon, wenn auch im Gegensatz zu CANN & SADLIER (2017) wie auch *burrungandjii* nicht auf Unterart-, sondern auf Artniveau, wie es auch die Erstbeschreibungen beider Taxa vorsahen. Die Einstufung von *walloyarrina* und *burrungandjii* als Unterarten von *Chelodina* (*Macrochelodina*) *oblonga* im vorliegenden Band folgt bis zur genaueren Klärung der Verwandtschaftsbeziehungen dieser Taxa dem Konzept von CANN & SADLIER (2017). Die Populationen von *Chelodina* (*Macrochelodina*) *oblonga rugosa* von Neuguinea werden in der Literatur verschiedentlich unter der Bezeichnung „*Chelodina siebenrocki* WERNER, 1901" als eigenständige Art aufgeführt (z. B. RHODIN & MITTERMEIER 1976, RHODIN & GENORUPA 2000), die jedoch von GEORGES *et al.* (2002) aufgrund der geringen genetischen Unterschiede lediglich als Synonym von *oblonga* angesehen wird (FRITZ & HAVAŠ 2006, 2007, TURTLE TAXONOMY WORKING GROUP 2007, 2009, 2010, 2011, 2012, 2014, 2017, RHODIN *et al.* 2008, GEORGES & THOMSON 2010).

Während in manchen der oben genannten Fälle fraglich ist, ob es sich bei den genannten Taxa tatsächlich um eigenständige Arten (oder Unterarten) handelt, ist dies bei der im Südwesten Australiens vorkommenden Form der Schlangenhalsschildkröten völlig unstreitig. Alles andere als unumstritten war jedoch in den vergangenen Jahren die wissenschaftliche Bezeichnung, die dieses Taxon korrekterweise zu tragen hat, wie auch der Diskussion oben bei *Chelodina* (*Macrochelodina*) *oblonga* zu entnehmen ist. Um den lange Jahre für diese Form gebräuchlichen, aber tatsächlich auf einem Exemplar der Nordaustralischen Schlangenhalsschildkröte begründeten Artnamen *oblonga* für die südwestliche Art zu erhalten, was auch SAVAGE (2007) vorschlägt, bestimmen McCORD & JOSEPH-OUNI (2007b) das von THOMSON (2000) nachträglich festgelegte Typusexemplar (Lectotypus) von *Chelodina colliei*, dem nach damaliger Einschätzung nächstälteren Namen für dieses Taxon, zum gleichzeitig neuen Typusexemplar (Neotypus) von *Chelodina oblonga*, für die sie in der gleichen Publikation den Gattungsnamen *Macrodiremys* aufstellen. Auf diese Weise wäre „*Chelodina oblonga*", mit einem nunmehr geografisch „passenden" neuen Typusexemplar, als Bezeichnung für die südwestaustralischen Schlangenhalsschildkröten sozusagen zu „retten" gewesen. Die Festlegung des Neotypus durch McCORD & JOSEPH-OUNI (2007b) war allerdings ungültig, da sie nur durch die *Internationale Kommission für Zoologische Nomenklatur* hätte erfolgen dürfen, weil das ursprüngliche Typusexemplar (Holotypus) von *Chelodina oblonga* ja noch existierte und lediglich viele Jahre ein Irrtum bezüglich seiner geografischen Herkunft bestand (KUCHLING 2010). Wie oben now as before meant the one from southwestern Australia) with the lectotype of *Chelodina colliei*, which they considered a junior synonym, that had subsequently been designated by THOMSON (2000), they had, from a taxonomic perspective, in fact created a new, entirely independent species name that immediately became a junior synonym of *Chelodina colliei*... The genus name *Macrodiremys* that was now based on the accidentally created junior synonym "*Macrodiremys oblonga*" by MCCORD & JOSEPH-OUNI (2007b) may therefore be applied to the older denomination for the same taxon, *Chelodina colliei*, upon which no genus name had been based before (TURTLE TAXONOMY WORKING GROUP 2014). According to GEORGES & THOMSON (2010), the cladistic relationships within the group of the side-necked turtles with long necks from Australia, New Guinea, and adjacent islands are best reflected by arranging all species in one genus, *Chelodina*, with altogether three subgenera, one of which is *Macrodiremys* (see above; TURTLE TAXONOMY WORKING GROUP 2014, 2017).

Genus *Elseya*:
The subdivision of this genus into subgenera and species here follows THOMSON *et al.* (2015). HOSER (2018) proposed another subgenus, "*Fitzroychelys*", for *Elseya albagula*, but it is not recognised here for the reasons outlined above.

In *Elseya* (*Elseya*) *dentata*, it may be expected that this name covers several, as yet undescribed forms (GEORGES & THOMSON 2010), including those occurring in the Kimberley region, the Fitzroy and Dawson Rivers, as well from the Victoria and Daly Rivers.

The species currently referred to as *Elseya* (*Elseya*) *flaviventralis* was first mentioned as a possibly distinct taxon by LEGLER (1981), has since been confirmed as such by various authors on the basis of morphological and genetic differentiations (see THOMSON & GEORGES 2016), and was described as "*Elseya flaviventralis*" by GEORGES *et al.* (2000) and under the denomination "*Elseya jukesi*" by WELLS (2002b). These two descriptions did not, however, formally conform to the minimum requirements of the *International Code of Zoological Nomenclature*, for which reason they represent *nomina nuda* (FRITZ & HAVAŠ 2006, TURTLE TAXONOMY WORKING GROUP 2017). Against this background, WELLS (2007b), like in other instances of the same kind, simply tried to redescribe this form under the same name, this time adhering to the rules of nomenclature. "*Elseya jukesi* WELLS, 2007" was then treated as valid in systematic checklists and more generalized herpetological literature for several years (TURTLE TAXONOMY WORKING GROUP 2007, 2009, RHODIN *et al.* 2008, WILSON & SWAN 2008, CANN & SADLIER 2017), but this name

bereits erläutert, entschied sich die INTERNATIONAL COMMISSION ON ZOOLOGICAL NOMENCLATURE (2013) schließlich dafür, dem Prinzip der Priorität zu folgen und den Namen *oblonga* für die nordaustralische Art festzuschreiben, weshalb für das südwestaustralische Taxon der nun älteste verfügbare Name *colliei* verwendet werden muss. Und spätestens jetzt wird es endgültig kompliziert. Indem MCCORD & JOSEPH-OUNI (2007b) die von ihnen „*Macrodiremys oblonga*" genannte Art (sie meinten damit nach wie vor jene aus dem Südwesten Australiens) mit dem von THOMSON (2000) nachträglich festgelegten Lectotypus des nach ihrer Auffassung jüngeren Synonyms *Chelodina colliei* in Verbindung setzen, schaffen sie aus taxonomischer Sicht damit eine neue, völlig eigenständige Artbezeichnung, die wiederum gleichzeitig ein jüngeres Synonym von *Chelodina colliei* darstellt... Der von MCCORD & JOSEPH-OUNI (2007b) auf dem nunmehr von ihnen unabsichtlich geschaffenen jüngeren Synonym „*Macrodiremys oblonga*" begründete Gattungsname *Macrodiremys* kann daher ebenso für die ältere Bezeichnung für das gleiche Taxon, *Chelodina colliei*, verwendet werden, auf der zuvor noch kein Gattungsname begründet wurde (TURTLE TAXONOMY WORKING GROUP 2014). Nach GEORGES & THOMSON (2010) werden die verwandtschaftlichen Verhältnisse innerhalb der Gruppe der langhalsigen Schlangenhalsschildkröten Australiens, Neuguineas und der benachbarten Inseln jedoch am besten dargestellt, indem alle Arten in eine Gattung *Chelodina* mit insgesamt drei Untergattungen gestellt werden, von denen *Macrodiremys* eine darstellt (s. o.; TURTLE TAXONOMY WORKING GROUP 2014, 2017).

Gattung *Elseya*:
Die Einteilung der Gattung in Untergattungen und Arten folgt THOMSON *et al.* (2015). Hoser (2018) stellt für *Elseya albagula* eine weitere Untergattung „*Fitzroychelys*" auf, die aus den oben genannten Gründen hier nicht anerkannt wird.

Bei *Elseya* (*Elseya*) *dentata* ist zu erwarten, dass sich unter dieser Bezeichnung mehrere noch unbeschriebene Formen verbergen (GEORGES & THOMSON 2010), so etwa im Bereich der Kimberley-Region, von Fitzroy River und Dawson River sowie von Victoria River und Daly River.

Die aktuell unter der Bezeichnung *Elseya* (*Elseya*) *flaviventralis* geführte Art wurde erstmals von LEGLER (1981) als möglicherweise eigenständiges Taxon erwähnt, von verschiedenen Autoren auf morphologischer und genetischer Basis als solches bestätigt (siehe THOMSON & GEORGES 2016) und bereits von GEORGES *et al.* (2000) als „*Elseya flaviventralis*" und von WELLS (2002b) unter dem Namen „*Elseya jukesi*" beschrieben. Diese beiden Beschreibungen genügen jedoch formell nicht den Anforderungen der *Internationalen Regeln für Zoologische Nomenklatur*, weshalb die beiden genannten Namen jeweils ein *nomen nudum* darstellen (FRITZ & HAVAŠ 2006, TURTLE TAXONOMY WORKING GROUP 2017). Aus diesem Grund unternimmt WELLS (2007b) wie auch in anderen Fällen den Versuch, die Form unter dem gleichen Namen erneut zu beschreiben, dieses Mal unter Beachtung der nomenklatorischen Regeln. „*Elseya jukesi* WELLS, 2007" wurde daraufhin jahrelang in systematische Checklisten und allgemein-herpetologische Literatur übernommen (TURTLE is actually not available for taxonomical purposes either due to its having been published in the *Australian Biodiversity Record* (see above; Fritz & Havaš 2007, Thomson & Georges 2009, Georges & Thomson 2010, Turtle Taxonomy Working Group 2010, 2011, 2012, 2014, 2017, Thomson & Georges 2016). This species was also, but this time properly, redescribed as "*Elseya flaviventralis*" by Thomson & Georges (2016), which name and authorship must now be considered the only valid one.

Various populations that are at present still assigned to *Elseya* (*Hanwarachelys*) *novaeguineae* and *Elseya* (*Hanwarachelys*) *schultzei*, respectively, possibly represent as yet undescribed species (see, e.g., Cann & Sadlier 2017 for figures). The populations of *Elseya* (*Pelocomastes*) *albagula* from the Fitzroy River and from the Burnett and Mary Rivers, respectively, may likewise in fact be species in their own rights (Todd *et al.* 2013). The Fitzroy population was described as a separate subspecies under the denomination "*Elseya albagula fitzroyi*" by Hoser (2018), but what was said in general about Hoser's works in the chapter for the genus *Pelochelys* also applies to this case.

Following Georges & Thomson (2010) and the Turtle Taxonomy Working Group (2010, 2011, 2012, 2014, 2017), *Elseya* (*Pelocomastes*) *irwini* here is treated as including the population in the Johnstone Rivers in northern Queensland, which was described as "*Elseya stirlingi*" by Wells & Wellington (1985) and referred to under this name in some recent publications, even if in part as a synonym of *Elseya dentata* (Fritz & Havaš 2006, 2007) or as the "Johnstone form" without a scientific name (Georges & Adams 1996). According to Iverson *et al.* (2001), the description of "*Elseya stirlingi*" does not conform to the requirements of the *International Code of Zoological Nomenclature*, so that this denomination is another unavailable *nomen nudum*. Moreover are *irwini* and "*stirlingi*" barely distinct from both a genetic and a morphological perspective so that distinguishing between them at species level does not appear warranted even though the two forms together take an isolated position within the genus *Elseya* (Georges & Thomson 2010). Wells (2007b) tried to eliminate the nomenclatural deficiencies of the original description by redescribing "*Elseya stirlingi*" and initially succeeded insofar as this name was temporarily included in systematic checklists and more generalized herpetological literature (Turtle Taxonomy Working Group 2007, 2009, Rhodin *et al.* 2008, Wilson & Swan 2008, Cann & Sadlier 2017), but this work was once more published in the repeatedly mentioned, controversial online journal *Australian Biodiversity Record* and is therefore regarded as unavailable for nomenclatural purposes (see above; Fritz & Havaš 2007, Thomson & Georges 2009, Georges & Thomson 2010, Turtle Taxonomy Working Group 2010, 2011, 2012, 2014, 2017). The populations in the Daintree River, which is likewise situated in northern Queensland, are closely related to *Elseya* (*Pelocomastes*) *irwini*, but may apparently represent an as yet undescribed species (Todd *et al.* 2014a, Cann & Sadlier 2017). It was in fact described as such as "*Elseya shireenhoserae*" by Hoser (2018), but this name is here ignored for the reasons given above as well.

Taxonomy Working Group 2007, 2009, Rhodin *et al.* 2008, Wilson & Swan 2008, Cann & Sadlier 2017), doch ist diese Bezeichnung aufgrund ihrer Veröffentlichung im *Australian Biodiversity Record* zu nomenklatorischen Zwecken nicht verfügbar (s. o.) (Fritz & Havaš 2007, Thomson & Georges 2009, Georges & Thomson 2010, Turtle Taxonomy Working Group 2010, 2011, 2012, 2014, 2017, Thomson & Georges 2016). Die ebenfalls wiederholte Beschreibung der Art als „*Elseya flaviventralis*" durch Thomson & Georges (2016) ist daher als die gültige anzusehen.

Verschiedene Populationen, die derzeit noch zu *Elseya* (*Hanwarachelys*) *novaeguineae* und *Elseya* (*Hanwarachelys*) *schultzei* gerechnet werden, stellen möglicherweise noch unbeschriebene Arten dar (Abbildungen siehe z. B. bei Cann & Sadlier 2017).

Auch die zu *Elseya* (*Pelocomastes*) *albagula* gestellten Populationen präsentieren evtl. jeweils eigenständige Arten, die im Fitzroy River bzw. in den Burnett und Mary Rivers verbreitet sind (Todd *et al.* 2013). Die Fitzroy-Population wird von Hoser (2018) als Unterart mit der Bezeichnung „*Elseya albagula fitzroyi*" beschrieben, doch gilt auch in diesem Fall das bereits im Kapitel zur Gattung *Pelochelys* allgemein zu den Arbeiten Hosers Geschriebene.

Zu *Elseya* (*Pelocomastes*) *irwini* wird hier entsprechend Georges & Thomson (2010) und der Turtle Taxonomy Working Group (2010, 2011, 2012, 2014, 2017) auch die Population in den Johnstone Rivers im nördlichen Queensland gerechnet, die von Wells & Wellington (1985) als „*Elseya stirlingi*" beschrieben und unter diesem Namen auch in der jüngeren Literatur geführt wurde, z. T. allerdings auch als Synonym von *Elseya dentata* (Fritz & Havaš 2006, 2007) oder als „Johnstone-Form" ohne wissenschaftlichen Namen (Georges & Adams 1996). Die Beschreibung von „*Elseya stirlingi*" genügt jedoch nach Iverson *et al.* (2001) nicht den Anforderungen der *Internationalen Regeln für Zoologische Nomenklatur*, so dass es sich bei dieser Bezeichnung um ein nicht verfügbares *nomen nudum* handelt. Zudem unterscheiden sich *irwini* und „*stirlingi*" in genetischer und morphologischer Hinsicht nur wenig voneinander, so dass eine Unterscheidung beider Formen auf Artebene nicht angebracht erscheint, auch wenn beide gemeinsam innerhalb der Gattung *Elseya* eine isolierte Position einnehmen (Georges & Thomson 2010). Wells (2007b) versucht zwar, durch eine erneute Beschreibung von „*Elseya stirlingi*" die nomenklatorischen Mängel der Erstbeschreibung zu beseitigen, was dem Namen die kurzfristige Aufnahme in systematische Checklisten und allgemein-herpetologische Literatur bescherte (Turtle Taxonomy Working Group 2007, 2009, Rhodin *et al.* 2008, Wilson & Swan 2008, Cann & Sadlier 2017), doch wurde auch diese Arbeit wieder in der bereits mehrfach erwähnten umstrittenen Onlinezeitschrift *Australian Biodiversity Record* veröffentlicht und gilt daher als zu nomenklatorischen Zwecken nicht verfügbar (s. o.) (Fritz & Havaš 2007, Thomson & Georges 2009, Georges & Thomson 2010, Turtle Taxonomy Working Group 2010, 2011, 2012, 2014, 2017). Die Populationen im Daintree River, ebenfalls im Norden Queenslands gelegen, stehen *Elseya* (*Pelocomastes*) *irwini* nahe, stellen aber offenbar eine noch unbeschriebene Art dar (Todd *et al.* 2014a, Cann & Sadlier

Genus *Emydura*:
Uncertainty continues to prevail regarding the true identity of the populations of *Emydura* native to the Australian north. Whereas some authors (e.g., WORRELL 1970, ERNST & BARBOUR 1989) supposed there was only one species (*Emydura australis*) in this region, others distinguished between a "red-faced" and a "yellow-faced" form that they referred to as *Emydura australis* and *Emydura victoriae*, respectively (e.g., COGGER 1975, CANN 1978) or the other way around (e.g., COGGER 2000). Like CANN (1998) before them, FRITZ & HAVAŠ (2006, 2007), the TURTLE TAXONOMY WORKING GROUP (2007, 2009), WILSON & SWAN (2008), and RHODIN *et al.* (2008) considered *Emydura australis* a separate species. Various other authors applied the name to populations in the Kimberley region of Western Australia (TUCKER *et al.* 2005, TKACH & SNYDER 2008), and CANN (1998) proved the existence of conspicuously small forms from the King Edward and the Prince Regent Rivers in the Kimberley region under this denomination, but it remains a fact that GRAY (1872) fixed the type locality of *australis* to the Macquarie River in New South Wales. Analysing the work by GRAY (1872), COGGER *et al.* (1983) supposed that *australis* was a junior synonym of *Emydura macquarii macquarii* that is native to the Macquarie River (see also GEORGES & THOMSON 2010, TURTLE TAXONOMY WORKING GROUP 2010, 2011, 2012, 2014, 2017, with the latter regarding *australis* as a *nomen dubium* due to the uncertain provenance of the type specimen). GEORGES & ADAMS (1992, 1996) presumed that there were three distinct species of the genus *Emydura* in northern Australia, which would even occur together in the river system of the Daly River: *Emydura victoriae*, *Emydura tanybaraga*, and *Emydura worrelli*. This scenario is also reflected in the taxonomic lists published by GEORGES & THOMSON (2010) and the TURTLE TAXONOMY WORKING GROUP (2010, 2011, 2012, 2014, 2017). On the other hand, CANN & SADLIER (2017) examined the type specimen of *Emydura australis* and ruled out the possibility that it was a representative of *Emydura macquarii* or would at all be referable to the eastern Australian members of this genus. They also argued, like several other authors mentioned above before, that this taxon was indigenous to the Kimberley region of Western Australia. Until more research results may paint a different picture, I here follow the concept suggested by CANN & SADLIER (2017) and suppose that there are five species of *Emydura* in northern Australia (*australis*, *subglobosa*, *tanybaraga*, *victoriae*, and *worrelli*).
The systematics of the genus *Emydura* in eastern Australia received the attention of GEORGES & ADAMS (1996), who could not detect genetic or morphological differences between the taxa *macquarii*, *krefftii* and *signata* that used to be treated as separate species in many publications before. They relegated the two latter names to the synonymy of *macquarii*, after GEORGES (1994) had already referred *signata* as a synonym to *krefftii*. Later CANN (1998) (*binjing*, *dharra*, *dharuk*, *gunabarra*), CANN *et al.* (2003) (*emmotti*), and McCORD *et al.*

2017); diese wird von HOSER (2018) als „*Elseya shireenhoserae*" beschrieben, was hier jedoch ebenfalls aus den bereits genannten Gründen nicht anerkannt wird.

Gattung *Emydura*:
Bezüglich der Identität der im Norden Australiens beheimateten *Emydura*-Populationen besteht bis heute Unsicherheit. Während manche Autoren (z. B. WORRELL 1970, ERNST & BARBOUR 1989) von der Existenz lediglich einer Art (*Emydura australis*) in dieser Region ausgehen, unterscheiden andere zwischen einer „rotgesichtigen" und einer „gelbgesichtigen" Form und nennen diese *Emydura australis* und *Emydura victoriae* (z. B. COGGER 1975, CANN 1978) bzw. umgekehrt (z. B. COGGER 2000). Auch CANN (1998), FRITZ & HAVAŠ (2006, 2007), die TURTLE TAXONOMY WORKING GROUP (2007, 2009), WILSON & SWAN (2008) und RHODIN *et al.* (2008) führen *Emydura australis* als eigenständige Art mit auf. Von verschiedenen Autoren wird der Name auf Populationen in der Kimberley-Region von Western Australia bezogen (TUCKER *et al.* 2005, TKACH & SNYDER 2008), und CANN (1998) belegt ausdrücklich kleinwüchsige Formen aus dem King Edward River und dem Prince Regent River in der Kimberley-Region mit dieser Bezeichnung, obwohl GRAY (1872) die Typuslokalität von *australis* auf den Macquarie River in New South Wales festgelegt hatte. COGGER *et al.* (1983) gehen aufgrund der Arbeit von GRAY (1872) davon aus, dass es sich bei *australis* um ein jüngeres Synonym der im Macquarie River beheimateten *Emydura macquarii macquarii* handelt (siehe auch GEORGES & THOMSON 2010, TURTLE TAXONOMY WORKING GROUP 2010, 2011, 2012, 2014, 2017; Letztere bezeichnet *australis* aufgrund der ungeklärten Herkunft des Typusexemplares als *nomen dubium*). GEORGES & ADAMS (1992, 1996) vermuten, dass im Norden Australiens drei Arten der Gattung *Emydura* zu unterscheiden sind, die im Flusssystem des Daly River sogar gemeinsam vorkommen: *Emydura victoriae*, *Emydura tanybaraga* und *Emydura worrelli*. Dies spiegelt sich auch in den taxonomischen Listen von GEORGES & THOMSON (2010) und der TURTLE TAXONOMY WORKING GROUP (2010, 2011, 2012, 2014, 2017) wider. CANN & SADLIER (2017) schließen nach der Untersuchung des Typusexemplares von *Emydura australis* aus, dass es sich um ein Exemplar von *Emydura macquarii* handelt oder überhaupt den ostaustralischen Vertretern der Gattung zuzuordnen ist und argumentieren wie schon andere oben genannte Autoren, dass dieses Taxon in der Kimberley-Region Western Australias beheimatet ist. Bis zum Vorliegen weiterer Untersuchungen folge ich im vorliegenden Band dem Konzept von CANN & SADLIER (2017), die von der Existenz von insgesamt fünf *Emydura*-Arten im Norden Australiens ausgehen (*australis*, *subglobosa*, *tanybaraga*, *victoriae*, *worrelli*).
Mit der Systematik der Gattung *Emydura* im östlichen Australien beschäftigen sich GEORGES & ADAMS (1996), die keine genetischen oder morphologischen Unterschiede zwischen den bis dahin meist als eigenständige Arten geführten Taxa *macquarii*, *krefftii* und *signata* feststellen können und die letzteren beiden daher als Synonyme von *macquarii* einstufen, nachdem GEORGES (1994) zuvor bereits *signata* als Synonym zu *krefftii* stellt. Später beschreiben CANN (1998) (*binjing*, *dhar-*

(2003) (*nigra*) described various subspecies of *macquarii* that, together with *krefftii* and *signata*, were accepted as such by FRITZ & HAVAŠ (2006, 2007), the TURTLE TAXONOMY WORKING GROUP (2007, 2009), and RHODIN *et al.* (2008). In the present volume, I follow the most recent arrangement suggested by CANN & SADLIER (2017) who once more reverted to considering *Emydura krefftii* a species in its own right and assigned it two subspecies, *emmotti* and *nigra*, due to their sharing some morphological traits (including a light stripe behind the eye on both sides); the populations to the north of the Burdekin River are possibly referable to an as yet undescribed subspecies (TODD *et al.* 2014b). As a consequence of their being indistinguishable at genetic level, GEORGES & THOMSON (2010) placed *binjing, dharra, dharuk, gunabarra,* and *signata* in the synonymy of *Emydura macquarii macquarii*; they interpreted differences in the shape of the shell and body size as expressions of variability in the nominate subspecies. The TURTLE TAXONOMY WORKING GROUP (2010, 2011, 2012, 2014, 2017) has likewise been regarding the mentioned forms as synonyms of *Emydura macquarii macquarii*. I have here again opted for following the concept suggested by CANN & SADLIER (2017) who considered the morphological differences between *binjing, dharra, dharuk, gunabarra, macquarii,* and *signata* sufficient for differentiating between subspecies within *Emydura macquarii. Emydura krefftii emmotti, Emydura krefftii nigra, Emydura macquarii dharra,* and *Emydura macquarii gunabarra* had already been described by WELLS & WELLINGTON (1985) under the denominations "*Chelymys windorah*", "*Tropicochelymys insularis*", "*Chelymys cooki*", and "*Chelymys johncanni*", respectively, which, according to IVERSON *et al.* (2001), did not conform to the *International Code of Zoological Nomenclature*, so that all of these names constitute *nomina nuda* without any nomenclatural relevance. The population currently assigned to *Emydura macquarii signata* from the river systems of the Tweed and Richmond Rivers in northeastern New South Wales must probably be regarded as representing yet another separate subspecies (CANN & SADLIER 2017). Often mentioned as a "candidate" for another separate subspecies and provisionally assigned to *Emydura macquarii dharra* here, the population in the Bellinger River (e.g., COGGER *et al.* 1993) is probably based upon specimens that were originally introduced through human activity from neighbouring river systems (GEORGES *et al.* 2007), even though older museum specimens from this river well exhibit differences to *Emydura macquarii dharra* as well as to the neighbouring *Emydura macquarii binjing* (CANN & SADLIER 2017). According to the TURTLE TAXONOMY WORKING GROUP (2010, 2011, 2012, 2014, 2017), and in contrast to the assessment by COWAN (1969), *Emydura macquarii* was first described by GRAY already in 1830, and not only in 1831, as is commonly stated in the existing literature.

The cladistic relationships between the populations of *Emydura subglobosa* in Australia and New Guinea have so far not been investigated in grater depth, but they exhibit at least some minor chromatic differences (CANN & SADLIER 2017). Should these forms need to be differentiated at species or subspecies levels in the future, the denomination "*Tropicochelymys goodie*" coined by WELLS & WELLINGTON (1985) for the Australian population would be unavailable according

ra, dharuk, gunabarra) sowie CANN *et al.* (2003) (*emmotti*) und MCCORD *et al.* (2003) (*nigra*) verschiedene Unterarten von *macquarii*, die zusammen mit *krefftii* und *signata* von FRITZ & HAVAŠ (2006, 2007), der TURTLE TAXONOMY WORKING GROUP (2007, 2009) und RHODIN *et al.* (2008) auch als solche aufgeführt werden. Im vorliegenden Band folge ich dem aktuellen Konzept von CANN & SADLIER (2017), die erneut *Emydura krefftii* als eigenständige Art einstufen und ihr aufgrund morphologischer Gemeinsamkeiten (u. a. ein heller Streifen beidseitig hinter den Augen) die Unterarten *emmotti* und *nigra* zuordnen; die Populationen ab dem Burdekin River nordwärts sind möglicherweise einer weiteren, noch unbeschriebenen Unterart zuzurechnen (TODD *et al.* 2014b). GEORGES & THOMSON (2010) stellen *binjing, dharra, dharuk, gunabarra* und *signata* aufgrund mangelnder Unterscheidbarkeit auf genetischer Ebene in die Synonymie von *Emydura macquarii macquarii*; Unterschiede in Panzerform und Körpergröße interpretieren sie als Ausdruck der Variabilität der Nominatform. Die TURTLE TAXONOMY WORKING GROUP (2010, 2011, 2012, 2014, 2017) sieht die genannten Formen ebenfalls als Synonyme von *Emydura macquarii macquarii* an. Auch in diesem Fall folge ich hier dem Konzept von CANN & SADLIER (2017), die die morphologischen Unterschiede zwischen *binjing, dharra, dharuk, gunabarra, macquarii* und *signata* als ausreichend für eine Differenzierung von Unterarten innerhalb von *Emydura macquarii* ansehen. *Emydura krefftii emmotti, Emydura krefftii nigra, Emydura macquarii dharra* und *Emydura macquarii gunabarra* wurden bereits von WELLS & WELLINGTON (1985) unter den Bezeichnungen „*Chelymys windorah*", „*Tropicochelymys insularis*", „*Chelymys cooki*" bzw. „*Chelymys johncanni*" beschrieben, was allerdings nach IVERSON *et al.* (2001) nicht den *Internationalen Regeln für Zoologische Nomenklatur* entsprechend erfolgte, so dass es sich in allen Fällen wiederum jeweils um ein *nomen nudum* ohne nomenklatorische Relevanz handelt. Die derzeit *Emydura macquarii signata* zugerechnete Population in den Flusssystemen von Tweed River und Richmond River im Nordosten von New South Wales ist vermutlich als weitere eigenständige Unterart einzustufen (CANN & SADLIER 2017). Die oft ebenfalls als „Kandidat" für eine weitere eigenständige Unterart genannte und hier provisorisch *Emydura macquarii dharra* zugeordnete Population im Bellinger River (z. B. COGGER *et al.* 1993) basiert heute wahrscheinlich auf vom Menschen aus benachbarten Flusssystemen eingeschleppten Exemplaren (GEORGES *et al.* 2007), während ältere Museumsexemplare aus diesem Fluss durchaus sowohl Unterschiede zu *Emydura macquarii dharra* als auch zu der benachbarten *Emydura macquarii binjing* zeigen (CANN & SADLIER 2017). Laut der TURTLE TAXONOMY WORKING GROUP (2010, 2011, 2012, 2014, 2017) wurde *Emydura macquarii*, wenn man den Angaben von COWAN (1969) folgt, von GRAY bereits im Jahre 1830 beschrieben, und nicht erst 1831, wie meist in der Literatur angegeben.

Die Verwandtschaftsbeziehungen der Populationen von *Emydura subglobosa* in Australien und Neuguinea wurden noch nicht näher untersucht, doch sind durchaus leichte Färbungsunterschiede festzustellen (CANN & SADLIER 2017); sollte zukünftig eine Unterscheidung auf Art- oder Unterartebene erfolgen, ist die von WELLS & WELLINGTON (1985) für die australische Population geprägte Bezeichnung

to IVERSON *et al.* (2001), because this name was just another nomenclaturally irrelevant *nomen nudum*.
Emydura worrelli was recognised as potentially distinctive already by CANN (1972, 1978) and subsequently formally described as "*Tropicochelymys worrelli*" by WELLS & WELLINGTON (1985), even if with an extremely brief characterization of its diagnostic traits, which, according to IVERSON *et al.* (2001), is just sufficient to permit the nomenclatural use of this name. The denomination "*Tropicochelymys leichhardti*" that was proposed by WELLS & WELLINGTON (1985) at the same time, on the other hand, is, according to IVERSON (2001) for one yet another *nomen nudum* with no nomenclatural relevance, and for the other, the population in the Leichhardt River is not regarded as a separate taxon, at least not at present. Subsequent to 1985, *worrelli* was considered a synonym of *Emydura victoriae* by IVERSON (1992) and IVERSON *et al.* (2001), but a good species by CANN (1998). Later, GEORGES & THOMSON (2006) recognised it, in part based upon the genetic studies by GEORGES & ADAMS (1992, 1996), like GEORGES *et al.* (2000) before, as a subspecies of *Emydura subglobosa*. This arrangement was followed by FRITZ & HAVAŠ (2006, 2007), the TURTLE TAXONOMY WORKING GROUP (2007, 2009, 2010, 2011, 2012, 2014, 2017), and RHODIN *et al.* (2008). GEORGES *et al.* (2006) retained this taxon as a separate species, and these authors also mentioned that an undescribed form of *Emydura* existed in the Trans-Fly region in the Western Province of Papua-New Guinea that was most similar to *worrelli*. According to GEORGES & THOMSON (2010), *subglobosa* and *worrelli* (which they considered subspecies of *Emydura subglobosa*) are distinctly different as per their preferred habitats and morphologies, and these authors called for thorough morphological and genetic studies to clarify the status of these two taxa. In the present volume, I have opted for following the most recent assessment by CANN & SADLIER (2017), who regarded *subglobosa* and *worrelli* as species in their own rights.

Genus *Myuchelys*:
Two clades had for long been differentiated within the genus *Elseya* (e.g., LEGLER 1981, GEORGES & ADAMS 1992, 1996, GEORGES & THOMSON 2006), until WELLS (2007c) eventually introduced the genus name *Wollumbinia* (with the type species *latisternum*) for the cladistic group around "*Elseya*" *latisternum* which subsequently found entry into systematic checklists and more generalized herpetological literature (TURTLE TAXONOMY WORKING GROUP 2007, 2009, RHODIN *et al.* 2008, WILSON & SWAN 2008, COGGER 2014, CANN & SADLIER 2017). However, this work was again published in the controversial online journal *Australian Biodiversity Record* that has been mentioned repeatedly and must therefore be regarded as unavailable for nomenclatural purposes (see above; FRITZ & HAVAŠ 2007, THOMSON & GEORGES 2009, GEORGES & THOMSON 2010, TURTLE TAXONOMY WORKING GROUP 2010, 2011, 2012, 2014, 2017). With the generic status of the cladistic group around "*Elseya*"/"*Wollumbinia*" *latisternum* being largely undisputed, THOMSON & GEORGES (2009) defined it as a separate genus under the denomination *Myuchelys*.

„*Tropicochelymys goodei*" laut IVERSON *et al.* (2001) allerdings nicht verfügbar, da es sich wieder um ein nomenklatorisch irrelevantes *nomen nudum* handelt.
Emydura worrelli wurde bereits von CANN (1972, 1978) als möglicherweise eigenständig erkannt und dann von WELLS & WELLINGTON (1985) als „*Tropicochelymys worrelli*" formal beschrieben, zwar mit einer äußerst knappen Darstellung der charakteristischen Merkmale, die jedoch nach IVERSON *et al.* (2001) ausreicht, um eine nomenklatorische Verwendbarkeit des Namens zu erlauben; die gleichzeitig von WELLS & WELLINGTON (1985) aufgestellte Bezeichnung „*Tropicochelymys leichhardti*" ist dagegen nach IVERSON (2001) zum einen wiederum ein *nomen nudum* ohne nomenklatorische Relevanz, zum anderen wird die Population des Leichhardt River zumindest derzeit nicht als eigenständiges Taxon angesehen. Nach 1985 wird *worrelli* von IVERSON (1992) und IVERSON *et al.* (2001) als Synonym von *Emydura victoriae*, von CANN (1998) dagegen als eigenständige Art angesehen. GEORGES & THOMSON (2006) wiederum stufen die Form dann, z. T. basierend auf den genetischen Studien von GEORGES & ADAMS (1992, 1996), wie zuvor schon GEORGES *et al.* (2000) als Unterart von *Emydura subglobosa* ein. Dieser Auffassung folgen FRITZ & HAVAŠ (2006, 2007), die TURTLE TAXONOMY WORKING GROUP (2007, 2009, 2010, 2011, 2012, 2014, 2017) und RHODIN *et al.* (2008). GEORGES *et al.* (2006) führen das Taxon wieder als eigenständige Art; diese Autoren erwähnen auch, dass eine in der Trans-Fly-Region in der Western Province von Papua-Neuguinea verbreitete, noch unbeschriebene *Emydura*-Form am meisten *worrelli* ähneln soll. Nach GEORGES & THOMSON (2010) unterscheiden sich *subglobosa* und *worrelli* (die sie als Unterarten von *Emydura subglobosa* ansehen) deutlich in Lebensraumwahl und Morphologie, und diese Autoren plädieren für gründliche morphologische und genetische Untersuchungen, um den Status beider Taxa zu klären. Im vorliegenden Band folge ich dem aktuellen Werk von CANN & SADLIER (2017), die *subglobosa* und *worrelli* als eigenständige Arten einstufen.

Gattung *Myuchelys*:
Schon seit geraumer Zeit wurden innerhalb der Gattung *Elseya* zwei Verwandtschaftsgruppen unterschieden (z. B. LEGLER 1981, GEORGES & ADAMS 1992, 1996, GEORGES & THOMSON 2006), bis schließlich WELLS (2007c) für die Verwandtschaftsgruppe um „*Elseya*" *latisternum* die Gattungsbezeichnung *Wollumbinia* (mit der Typusart *latisternum*) aufstellte, die auch in systematische Checklisten und die allgemein-herpetologische Literatur Einzug hielt (TURTLE TAXONOMY WORKING GROUP 2007, 2009, RHODIN *et al.* 2008, WILSON & SWAN 2008, COGGER 2014, CANN & SADLIER 2017), doch wurde diese Arbeit wieder in der bereits erwähnten umstrittenen Onlinezeitschrift *Australian Biodiversity Record* veröffentlicht und gilt daher ebenfalls als zu nomenklatorischen Zwecken nicht verfügbar (s. o.) (FRITZ & HAVAŠ 2007, THOMSON & GEORGES 2009, GEORGES & THOMSON 2010, TURTLE TAXONOMY WORKING GROUP 2010, 2011, 2012, 2014, 2017). Da der Gattungsstatus der Verwandtschaftsgruppe um „*Elseya*"/*Wollumbinia*" *latisternum* jedoch weitestgehend unstreitig ist, beschreiben THOMSON & GEORGES (2009) die Gattung erneut unter der Bezeichnung *Myuchelys*.

The original description of *Myuchelys bellii* (GRAY, 1844) did not contain information on a type locality. It was only CANN (1998) who, on the basis of morphological traits, managed to associate this species name with the population in the river systems of the Namoi, Gwydir, Severn, and Deepwater Rivers. CANN (1998) regarded the population in the neighbouring Bald Rock Creek in southeastern Queensland as a separate species, which he did not formally describe, though. WELLS (2002b) then described this form under the denomination *"Elseya dorriani"*, but for one, this was another publication in the controversial *Australian Biodiversity Record* (see above), and for the other, WELLS failed to designate a type specimen for his new "species", in which case the *International Code of Zoological Nomenclature* stipulate that the name has to be considered a *nomen nudum* without any nomenclatural bearing (FRITZ & HAVAŠ 2006, TURTLE TAXONOMY WORKING GROUP 2007). WELLS (2007c) himself subsequently downgraded *"dorriani"* to a subspecies of *Myuchelys bellii*. More recent genetic analyses have since revealed that the differences between the populations from the river systems of the Namoi, Gwydir, Severn and Deepwater on the one, and those from the Bald Rock Creek on the other hand are really minor (FIELDER *et al.* 2012), and this fact and the absence of morphological data made GEORGES & THOMSON (2010) deem them members of one and the same species.

WELLS (2009) described the populations of *Myuchelys latisternum* from southeastern Queensland as a separate species, *"Wollumbinia dorsii"* (TURTLE TAXONOMY WORKING GROUP 2009). This is again a publication in the *Australian Biodiversity Record* that therefore constitutes a denomination that is unavailable for nomenclatural purposes (see above; GEORGES & THOMSON 2010, TURTLE TAXONOMY WORKING GROUP 2010, 2011, 2012, 2014, 2017). In the same work, WELLS (2009) considered the type specimen of *Euchelymys spinosa* (GRAY, 1871) that had been collected in an unspecified location ("North Australia") in 1866 (CANN 1998) so significantly different from *Myuchelys latisternum* that he granted it species status in the combination *"Wollumbinia spinosa"* as well. Whether this assessment is true to the facts may be questionable, especially as no further specimens have become known and the exact distribution range of this "species" has remained a mystery. These reasons were sufficient for the TURTLE TAXONOMY WORKING GROUP (2009, 2010, 2011, 2012, 2014, 2017) and CANN & SADLIER (2017) to regard *spinosa* as a synonym of *Myuchelys latisternum* at least for the time being. The variability of *Myuchelys latisternum* is at present a subject of scientific research (FREEMAN & CANN 2014)

In the framework of their genetic analyses of the species of the genera *Emydura*, *Elseya*, *Myuchelys*, *Elusor* and *Rheodytes*, LE *et al.* (2013) noted that the species *purvisi*, which had until then been assigned to the genus *Myuchelys*, was sister to the species of the genera *Emydura*, *Elseya* and *Myuchelys* and therefore placed it in a separate genus, *Flaviemys*. This result corroborated similar conclusions by GEORGES & ADAMS (1992), GEORGES *et al.* (1998), THOMSON & GEORGES (2009), GEORGES & THOMSON (2010), FIELDER *et al.* (2012), and FIELDER (2013), but was challenged once more by SPINKS *et al.* (2015) so that in the present volume *purvisi* is assigned to

Bei der Erstbeschreibung von *Myuchelys bellii* legt GRAY (1844) keine Typuslokalität fest. Erst CANN (1998) ordnet anhand morphologischer Merkmale diesen Artnamen der Population in den Flusssystemen von Namoi, Gwydir, Severn und Deepwater River zu. CANN (1998) betrachtet die Population im benachbarten Bald Rock Creek im Südosten Queenslands als eigenständige Art, ohne sie allerdings formell zu beschreiben. WELLS (2002b) beschreibt diese Form unter der Bezeichnung *„Elseya dorriani"*. Zum einen handelt es sich jedoch erneut um eine Veröffentlichung im umstrittenen *Australian Biodiversity Record* (s. o.), und zum anderen versäumt es WELLS, ein Typusexemplar für die neue „Art" zu bestimmen, so dass es sich nach den *Internationalen Regeln für Zoologische Nomenklatur* schon aus diesem Grund um ein *nomen nudum* ohne nomenklatorische Relevanz handelt (FRITZ & HAVAŠ 2006, TURTLE TAXONOMY WORKING GROUP 2007). WELLS (2007c) selbst stuft *„dorriani"* später als Unterart von *Myuchelys bellii* ein. Neuere genetische Untersuchungen zeigen, dass die Unterschiede zwischen den Populationen aus den Flusssystemen von Namoi, Gwydir, Severn und Deepwater River einerseits und dem Bald Rock Creek andererseits nur gering sind (FIELDER *et al.* 2012), und angesichts dieser Tatsache und des Fehlens morphologischer Daten betrachten GEORGES & THOMSON (2010) beide Formen als Angehörige einer Art.

WELLS (2009) beschreibt die Populationen von *Myuchelys latisternum* im Südosten Queenslands als eigenständige Art, *„Wollumbinia dorsii"* (TURTLE TAXONOMY WORKING GROUP 2009). Erneut handelt es sich jedoch um eine Veröffentlichung im *Australian Biodiversity Record* und somit um eine nicht zu nomenklatorischen Zwecken verfügbare Bezeichnung (s. o.) (GEORGES & THOMSON 2010, TURTLE TAXONOMY WORKING GROUP 2010, 2011, 2012, 2014, 2017). In der gleichen Arbeit sieht WELLS (2009) das im Jahre 1866 (CANN 1998) an einem nicht näher bekannten Ort („North Australia") gefangene Typusexemplar von *Euchelymys spinosa* (GRAY 1871) als so stark von *Myuchelys latisternum* abweichend an, dass er dieser Form unter der Bezeichnung *„Wollumbinia spinosa"* ebenfalls Artstatus zugesteht. Ob diese Einschätzung den Tatsachen entspricht, bleibt fraglich, zumal keine weiteren Exemplare bekannt sind und das genaue Verbreitungsgebiet der „Art" unbekannt ist. Aus diesem Grunde sieht die TURTLE TAXONOMY WORKING GROUP (2009, 2010, 2011, 2012, 2014, 2017) *spinosa* ebenso wie CANN & SADLIER (2017) zumindest vorerst weiterhin als Synonym von *Myuchelys latisternum* an. Die Variabilität von *Myuchelys latisternum* ist derzeit Gegenstand wissenschaftlicher Untersuchungen (FREEMAN & CANN 2014)

LE *et al.* (2013) stellen im Rahmen ihrer genetischen Untersuchung an den Arten der Gattungen *Emydura*, *Elseya*, *Myuchelys*, *Elusor* und *Rheodytes* fest, dass die bis dahin zur Gattung *Myuchelys* gestellte Art *purvisi* in einem Schwestergruppenverhältnis zu den Arten der Gattungen *Emydura*, *Elseya* und *Myuchelys* steht und stellen diese Art daher in eine eigenständige Gattung *Flaviemys*. Dieses Studienergebnis bestätigt ähnliche Schlussfolgerungen von GEORGES & ADAMS (1992), GEORGES *et al.* (1998), THOMSON & GEORGES (2009), GEORGES & THOMSON (2010), FIELDER *et al.* (2012) und FIELDER (2013), wird von SPINKS *et al.* (2015) allerdings the genus *Myuchelys*, which is in agreement with the Turtle Taxonomy Working Group (2017) and Cann & Sadlier (2017).

How to use this book

Pointer guidelines for the captive husbandry of the figured species and subspecies can be found in the fold-out leaf at the end of this book. The following general information is more adequately placed here, preceding the illustrative section.

All species and subspecies are by far best kept separated by their sexes outside of the mating season, even in the case of rather inaggressive turtles, as breeding results will usually be much more prolific if males and females are placed together only temporarily. Only a small number of species appear to require the continuous presence of the other sex for reproducing successfully. Considering that many keepers of these turtles are simply unable to provide the very large installations that are necessary to keep specimens apart or together, as may be the case, I have limited recommendations such as "keep sexes separated" or "keep individually" to those species for which in-depth contemplation suggested that such husbandry management would be crucial to keeping the animals healthy and essential for enabling them to reproduce on a regular basis.

Insofar as information on the dimensions of enclosures is concerned, the latter should be regarded as what they are called, that is, as **minimum requirements**. With this book catering mainly, but of course by no means exclusively, for a German-speaking audience, recommendations with regard to outdoor stays are based on the Central European climate. In closing it must be pointed out that the husbandry practices recommended here are of course **not** meant to substitute the detailed study of subject literature on the husbandry of these turtles.

Abbreviations used in the captions for Australian states are:

WA – Western Australia
NT – Northern Territory
QLD – Queensland
NSW – New South Wales
VIC – Victoria

Museum Acronyms used in this book:
WAM – Western Australian Museum (Australia)
SMF – Senckenberg Forschungsinstitut und Naturmuseum (Germany)
ZSM – Zoologische Staatssammlung München (Germany)

wieder in Frage gestellt, so dass *purvisi* im vorliegenden Band in Übereinstimmung mit der Turtle Taxonomy Working Group (2017) und Cann & Sadlier (2017) zur Gattung *Myuchelys* gestellt wird.

Benutzerhinweise

Richtlinien zur Haltung der abgebildeten Arten und Unterarten sind dem Faltblatt am Ende des Buches zu entnehmen. Folgende allgemeine Hinweise seien jedoch an dieser Stelle dem Bildteil vorangestellt.

Absolut erstrebenswert ist bei allen Arten eine getrennte Haltung der Geschlechter außerhalb der Paarungszeit; selbst bei verträglichen Schildkrötenarten führt eine nur zeitweise Vergesellschaftung männlicher und weiblicher Tiere meist zu einer deutlichen Steigerung des Nachzuchterfolges. Nur wenige Arten benötigen offenbar die ständige Vergesellschaftung der Geschlechter für eine erfolgreiche Fortpflanzung. Da aber vielen Schildkrötenhaltern eine getrennte Haltung oder die Unterbringung der Tiere in sehr weitläufigen Anlagen schlichtweg nicht möglich ist, habe ich hier die Empfehlung „Geschlechtertrennung" oder gar „Einzelhaltung" nur für solche Arten übernommen, bei denen ich nach sorgfältigem Abwägen zu dem Ergebnis gekommen bin, dass diese Art der Haltung für die Gesundheit der Tiere oder den regelmäßigen Zuchterfolg absolut notwendig ist.

Was die Angaben zur Behältergröße betrifft, so sind sie als das zu betrachten, als was sie auch bezeichnet werden, nämlich als **Mindestanforderungen**. Da dieses Buch sich überwiegend (wenn auch natürlich nicht ausschließlich) an die deutschsprachige Leserschaft wendet, beziehen sich auch die Empfehlungen zur Freilandhaltung auf mitteleuropäische Klimaverhältnisse. Abschließend sei noch bemerkt, dass die hier aufgeführten Haltungsempfehlungen selbstverständlich **nicht** das Studium weiterführender Literatur zur Pflege der Schildkröten ersetzen können und wollen.

In Bildunterschriften werden folgenden **Abkürzungen** für australische Bundesstaaten verwendet:

WA – Western Australia
NT – Northern Territory
QLD – Queensland
NSW – New South Wales
VIC – Victoria

Abkürzungen von Museen, die verwendet wurden:
WAM – Western Australian Museum (Australien)
SMF – Senckenberg Forschungsinstitut und Naturmuseum (Deutschland)
ZSM – Zoologische Staatssammlung München (Deutschland)

References/Literaturverzeichnis

The following list of references contains exclusively those treatises that have been referred to in the Introduction. A compilation of further reading material has been done without in a quest to save as much space as possible for the photograph section.

Das folgende Literaturverzeichnis nennt ausschließlich Arbeiten, auf die in der Einführung Bezug genommen wird. Zungunsten des Bildteils wird auf eine Zusammenstellung weiterführender Literatur verzichtet.

ALACS, A. A. (2008): Forensics, Phylogeography and Population Genetics: A Case Study using the Northern Snake-necked turtle, *Chelodina rugosa*. – PhD Thesis, University of Canberra, Canberra, 213 pp.

BATSCH, A. J. G. C. (1788): Versuch einer Anleitung, zur Kenntniß und Geschichte der Thiere und Mineralien, für akademische Vorlesungen entworfen, und mit den nöthigsten Abbildungen versehen. Erster Theil. Allgemeine Geschichte der Natur; besondre der Säugthiere, Vögel, Amphibien und Fische. Jena (Akademische Buchhandlung), 528 pp.

BECK, R. G. (1991): The Common Long-Necked Tortoise *Chelodina longicollis* (SHAW 1802) (Testudinata: Chelidae): A comparative study of the morphology and behaviour of disjunct populations. – South Australian Naturalist, **66** (1): 4–22.

BELL, T. (1828): On *Hydraspis*, a new genus of freshwater tortoises, of the family Emydidae. – Zoological Journal, London, **3** (12): 511–513.

BHULLAR, B. S. & G. S. BEVER (2009): An archosaur-like laterosphenoid in early turtles (Reptilia: Pantestudines). – Breviora, Cambridge, **518**: 1–11.

BONAPARTE, C. L. (1836): Cheloniorum Tabula Analytica. – Rome, 9 pp.

BOWEN, B. W. & S. A. KARL (2007): Population genetics and phylogeography of sea turtles. – Molecular Ecology, **16** (23): 4886–4907.

BOWEN, B. W., A. B. MEYLAN, J. P. ROSS, C. J. LIMPUS, G. H. BALAZS & J. C. AVISE (1992): Global population structure and natural history of the green turtle (*Chelonia mydas*) in terms of matriarchal phylogeny. – Evolution, **46** (4): 865–881.

BOWEN, B. W., W. S. NELSON & J. C. AVISE (1993): A molecular phylogeny for marine turtles: trait mapping, rate assessment, and conservation relevance. – Proceedings of the National Academy of Science, Washington, **90** (12): 5574–5577.

CANN, J. (1972): Notes on some tortoises collected in northern Australia. – Victorian Naturalist, Blackburn, **89**: 165–168.

CANN, J. (1978): Tortoises of Australia. – Sydney (Angus and Robertson), 79 pp.

CANN, J. (1998): Australian Freshwater Turtles. – Singapore (Beaumont Publishing), 292 pp.

CANN, J., W. P. MCCORD & M. JOSEPH-OUNI (2003): Emmott's shortneck turtle, *Emydura macquarii emmotti* ssp. nov. Pp. 60–61. – *In*: MCCORD, W. P., J. CANN & M. JOSEPH-OUNI (Eds.): A taxonomic assessment of *Emydura* (Testudines: Chelidae) with descriptions of new subspecies from Queensland, Australia. – Reptilia (GB), Castelldefels, **27**: 59–63.

CANN, J. & R. SADLIER (2017): Freshwater Turtles of Australia. – Rodeo (ECO Wear & Publishing), 448 pp.

CHESSMAN, B. C. (1978): Ecological studies of freshwater turtles in south-eastern Australia. – Ph.D. Thesis, Monash University, Melbourne, 428 pp.

COGGER, H. G. (1975): Reptiles and Amphibians of Australia. – Sydney (Angus and Robertson), 584 pp.

COGGER, H. G. (2000): Reptiles and Amphibians of Australia. 6th revised edition. – Sydney, Auckland, London & Cape Town (Reed New Holland), 808 pp.

COGGER, H. G. (2014): Reptiles and Amphibians of Australia. 7th revised edition. – Collingwood (CSIRO Publishing), 1.033 pp.

COGGER, H. G., E. E. CAMERON & H. M. COGGER (1983): Zoological Catalogue of Australia. Amphibia, Reptilia. – Canberra (Australian Government Publishing Service), VI + 313 pp.

COGGER, H. G., E. CAMERON, R. SADLIER & P. EGGLER (1993): The Action Plan for Australian Reptiles. – Canberra (Australian Nature Conservation Agency), 254 pp.

COLLINS, J. T., S. L. COLLINS & T. W. TAGGART (2010): Amphibians, Reptiles, and Turtles in Kansas. – Eagle Mountain (Eagle Mountanin Publishing), 312 pp.

COWAN, C. F. (1969): Notes on GRIFFITH's *Animal Kingdom of Cuvier* (1824–1835). – Journal of the Society of Bibliography of Natural History, Edinburgh, **5** (2): 137–140.

CRAWFORD, N. G., B. C. FAIRCLOTH, J. E. MCCORMACK, R. T. BRUMFIELD, K. WINKER & T. C. GLENN (2012): More than 1000 ultraconserved elements provide evidence that turtles are the sister group of archosaurs. – Biology Letters, London, **8** (5): 783–786.

DEBRAGA, M. & O. RIEPPEL (1997): Reptile phylogeny and the interrelationships of turtles. – Zoological Journal of the Linnean Society, London, **120**: 281–354.

DERANIYAGALA, P. E. P. (1933): The loggerhead turtles (Carettidae) of Ceylon. – Ceylon Journal of Science, Peradeniya, 18B: 59–70.

DERANIYAGALA, P. E. P. (1952): The loggerhead turtles (Carettinae) of Europe. – Herpetologica **8** (3): 57–58.

DUBOIS, A. & R. BOUR (2010): The distinction between family-series and class-series nomina in zoological nomenclature, with emphasis on the nomina created by BATSCH (1788, 1789) and on the higher nomenclature of turtles. – Bonn zoological Bulletin, **57** (2): 149–171.

DUCHENE, S., A. FREY, A. ALFARO-NÚÑEZ, P. H. DUTTON & M. T. P. GILBERT (2012): Marine turtle mitogenome phylogenetics and evolution. – Molecular Phylogenetics and Evolution, **65** (1): 241–250.

DUTTON, P. H., S. E. RODEN, K. R. STEWART, E. LACASELLA, M. TIWARI, A. FORMIA, J. C. THOMÉ, S. R. LIVINGSTONE, S. ECKERT, D. CHACON-CHAVERRI, P. RIVALAN & P. ALLMAN (2013): Population stock structure of leatherback turtles (*Dermochelys coriacea*) in the Atlantic revealed using mtDNA and microsatellite markers. – Conservation Genetics, **14** (3): 625–636.

ELLIS, R. J. & A. GEORGES (2015): An annotated type catalogue of the turtles (Testudines: Pleurodira: Chelidae) in the collection of the Western Australian Museum. – Records of the Western Australian Museum, Perth, **30** (1): 52–60.

ERNST, C. H. & R. W. BARBOUR (1989): Turtles of the World. – Washington & London (Smithsonian Institution Press), 313 pp.

ESCHSCHOLTZ, J. F. VON (1829a): Beschreibungen dreier neuer Meerschildkröten. – Die Quatember, Jelgava, **1** (1):10–18.

ESCHSCHOLTZ, J. F. VON (1829b): Zoologischer Atlas, enthaltend Abbildungen und Beschreibungen neuer Thierarten, während des Flottcapitains VON KOTZEBUE zweiter Reise um die Welt, auf der

Russisch-Kaiserlichen Kriegsschlupp Predpriaetiё in den Jahren 1823–1826. Erstes Heft. – Berlin (G. Reimer), 17 pp. + 15 pl.

Fielder, D. (2013): Ancient Phenotypes Revealed Through Present Day Species – A Morphological Analysis of Australia's Saw-Shelled Turtles Including the Threatened *Myuchelys bellii* (Testudines: Chelidae). – Chelonian Conservation and Biology, Lawrence, **12** (1): 101–111.

Fielder, D., K. Vernes, E. Alacs & A. Georges (2012): Mitochondrial variation among Australian freshwater turtles (genus *Myuchelys*), with special reference to the endangered *M. bellii*. – Endangered Species Research, Oldendorf, **17**, 63–71.

Flores-Villela, O., K. Adler & T. G. Eimermacher (2016): Identity of three new sea turtles named by J. Friedrich Eschscholtz. – Chelonian Conservation and Biology, Lawrence, **15** (1): 157–162.

Freeman, A. & J. Cann (2014): *Myuchelys latisternum* (Gray 1867) – Sawshelled Turtle, Saw-Shell Turtle. – *In*: Rhodin, A. G. J., P. C. H. Pritchard, P. P. van Dijk, R. A. Saumure, K. A. Buhlmann, J. B. Iverson & R. A. Mittermeier (Eds.): Conservation Biology of Freshwater Turtles and Tortoises: A Compilation Project of the IUCN/SSC Tortoise and Freshwater Turtle Specialist Group. – Chelonian Research Monographs, Lunenburg, **5** (7): 073.1–073.8.

Fritz, U. & P. Havaš (2006): Checklist of Chelonians of the World at the request of the CITES Nomenclature Committee and the German Agency for Nature Conservation. – Dresden (German Federal Ministry of Environment, Nature Conservation and Nuclear Safety and Museum of Zoology Dresden), 230 pp.

Fritz, U. & P. Havaš (2007): Checklist of Chelonians of the World. – Vertebrate Zoology, Dresden, **57** (2): 149–368.

Gaffney, E. S. (1977): The side-necked turtle family Chelidae: a theory of relationships using shared derived characters. – American Museum Novitates, New York, **2620**: 1–28.

Gaos, A. R., R. L. Lewison, M. J. Liles, V. Gadea, E. Altamirano, A. V. Henríquez, P. Torres, J. Urteaga, F. Vallejo, A. Baquero, C. LeMarie, J. P. Muñoz, J. A. Chaves, C. E. Hart, A. Peña de Niz, D. Chácon, L. Fonseca, S. Otterstrom, I. L. Yañez, E. L. LaCasella, A. Frey, M. P. Jensen & P. H. Dutton (2016): Hawksbill turtle terra incognita: conservation genetics of eastern Pacific rookeries. – Ecology and Evolution, **6**: 1251–1264.

Georges, A. (1994): Setting conservation priorities for Australian freshwater turtles. Pp. 49–58. – *In*: Lunney, D. & D. Ayers (Eds.): Herpetology in Australia – A Diverse Discipline. – Transactions of the Royal Society of New South Wales, Mosman, 414 pp.

Georges, A. & M. Adams (1992): A phylogeny for Australian chelid turtles based on allozyme electrophoresis. – Australian Journal of Zoology, Collingwood, **40** (5): 453–476.

Georges, A. & M. Adams (1996): Electrophoretic delineation of species boundaries within the short-necked chelid turtles of Australia. – Zoological Journal of the Linnean Society, London, **118**, 241–260.

Georges, A. & S. Thomson (2006): Evolution and zoogeography of the Australian freshwater turtles. – *In*: Merrick, J. R., M. Archer, G. M. Hickey & M. S. Y. Lee (Eds): Evolution and Zoogeography of Australasian Vertebrates. – Sydney (Australian Scientific Publishing): 291–308.

Georges, A. & S. Thomson (2010): Diversity of Australasian freshwater turtles, with an annotated synonymy and keys to species. – Zootaxa, Auckland, **2496**: 1–37.

Georges, A., J. Birrell, K. M. Saint, W. P. McCord & S. C. Donnellan (1998): A phylogeny for side-necked turtles (Chelonia: Pleurodira) based on mitochondrial and nuclear gene sequence variation. – Biological Journal of the Linnean Society, London, **67** (2): 213–246.

Georges, A., J. S. Doody, J. Young & J. Cann (2000): The Australian Pig-Nosed Turtle. – Canberra (Robey), 37 pp.

Georges, A., M. Adams & W. P. McCord (2002): Electrophoretic delineation of species boundaries within the genus *Chelodina* (Testudines: Chelidae) of Australia, New Guinea and Indonesia. – Zoological Journal of the Linnean Society, London, **134** (4): 401–421.

Georges, A., F. Guarino & B. Bito (2006) Freshwater turtles of the TransFly Region of Papua New Guinea –Notes on diversity, distribution, reproduction, harvest and trade. – Wildlife Research, Collingwood, **33** (5): 373–384.

Georges, A., R. Walsh, R. J. Spencer, M. Welsh & H. B. Shaffer (2007): The Bellinger *Emydura* – Challenges for Management – Report to NSW National Parks and Wildlife Service, Sydney, by the Institute for Applied Ecology. – Canberra (University of Canberra), 20 pp.

Georges, A., J. S. Doody, C. Eisemberg, E. Alacs & M. Rose (2008): *Carettochelys insculpta* Ramsay 1886 – Pig-Nosed Turtle, Fly River Turtle. – *In*: Rhodin, A. G. J., P. C. H. Pritchard, P. P. van Dijk, R. A. Saumure, K. A. Buhlmann & J. B. Iverson (Eds.): Conservation Biology of Freshwater Turtles and Tortoises: A Compilation Project of the IUCN/SSC Tortoise and Freshwater Turtle Specialist Group. – Chelonian Research Monographs, Lunenburg, **5** (1): 009.1–009.17.

Gray, J. E. (1825): A synopsis of the genera of reptiles and amphibia, with a description of some new species. – Annals of Philosophy, London, **2** (10): 193–217.

Gray, J. E. (1830): A Synopsis of the Species of the Class Reptilia. – *In*: Griffith, E. & E. Pidgeon: The Class Reptilia arranged by the Baron Cuvier, with specific descriptions. – *In*: Griffith, E. (Ed.): The Animal Kingdom Arranged in Conformity with its Organization, by the Baron Cuvier, with Additional Descriptions of all the Species Hitherto Named, and of many not before Noticed. Vol. 9. Reptilia. Part 26. Supplement. – London (Whittaker, Treacher and Co.). 110 pp.

Gray, J. E. (1841): A catalogue of the species of reptiles and amphibia hitherto described as inhabiting Australia, with a description of some new species from Western Australia, and some remarks on their geographical distribution. – *In*: Grey, G.: Journal of Two Expeditions of Discovery in Northwest and Western Australia. – London (T. and W. Boone), Volum 2, Appendix E: 422–449.

Gray, J .E. (1844): Catalogue of the tortoises, crocodiles and amphisbaenians in the collection of the British Museum. – London (Edward Newman), 80 pp.

Gray, J. E. (1871): On *Euchelymys*, a new genus and two new species of Australian freshwater tortoises. – Annals and Magazine of Natural History, London, Series 4, **8**: 117–118.

Gray, J. E. (1872) On the genus *Chelymys* and its allies from Australia. – Proceedings of the Zoological Society of London, London, **1872**: 504–514.

Guillon, J. M., L. Guéry, V. Hulin & M. Girondot (2012): A large phylogeny of turtles (Testudines) using molecular data. – Contributions to Zoology, Leiden, **81** (3): 147–158.

Hallermann, J. (2007): *„Deus creavit, Linnaeus disposuit"*. Zum 300. Geburtstag von Carl von Linné (1707–1778), dem Begründer der binominalen Nomenklatur in der Systematik. – Sekretär, Rheinbach, **7** (2): 3–8.

Hodges, K., S. Donnellan & A. Georges (2014): Phylogeography of the Australian freshwater turtle *Chelodina expansa* reveals complex relationships among inland and coastal bioregions. – Biological Journal of the Linnean Society, London, **111**: 789–805.

Hodges, K., S. Donnellan & A. Georges (2015): Significant genetic structure despite high vagility revealed through mitochondrial phylogeography of an Australian freshwater turtle (*Chelodina longicollis*). – Marine and Freshwater Research, Clayton, **66** (11): 1045–1056.

Hoser, R. T. (2014a): A review of the turtle genus *Pelochelys* Gray,

1864 (Trionychidae) including the division into two subgenera and the formal descriptions of two new species. – Australasian Journal of Herpetology, Park Orchards, **22**: 60–64.

HOSER, R. T. (2014b): A taxonomic revision of the Giant Long-necked Terrapin, *Chelodina expansa* GRAY, 1857 species complex and related matters of taxonomy and nomenclature. – Australasian Journal of Herpetology, Park Orchards, **24**: 3–11.

HOSER, R. T. (2018): A new subgenus, new species and new subspecies of *Elseya* GRAY, 1867 (Testudinata: Pleurodira: Chelidae) from Eastern Australia. – Australasian Journal of Herpetology, Park Orchards, **36**: 28–30.

INTERNATIONAL COMMISSION ON ZOOLOGICAL NOMENCLATURE (1999): International Code of Zoological Nomenclature. 4th revised edition. – London (The International Trust for Zoological Nomenclature, The Natural History Museum), 336 pp.

INTERNATIONAL COMMISSION ON ZOOLOGICAL NOMENCLATURE (2013): Opinion 2315 (Case 3351). *Chelodina rugosa* OGILBY, 1890 (currently *Macrochelodina rugosa*; Reptilia, Testudines): precedence not granted over *Chelodina oblonga* GRAY, 1841. – Bulletin of Zoological Nomenclature, London, **70** (1): 57–60.

IVERSON, J. B. (1992): A Revised Checklist with Distribution Maps of the Turtles of the World. – Richmond, (Privately Printed), XIII + 363 pp.

IVERSON, J. B, S. THOMSON & A. GEORGES (2001): Validity of the taxonomic changes for turtles proposed by WELLS and WELLINGTON. – Journal of Herpetology, **35** (3): 365–368.

KAISER, H., B. I. CROTHER, C. M. R. KELLY, L. LUISELLI, M. O'SHEA, H. OTA, P. PASSOS, W. D. SCHLEIP & W. WÜSTER (2013): Best practices: In the 21st century, taxonomic decisions in herpetology are acceptable only when supported by a body of evidence and published via peer-review. – Herpetological Review, **44** (1): 8–23.

KAISER, H. (2014): Best practices in herpetological taxonomy: errata and addenda. – Herpetological Review, **45** (2): 257–268.

KARL, S. A. & B. W. BOWEN (1999): Evolutionary significant units versus geopolitical taxonomy - molecular systematics of an endangered sea turtle (genus *Chelonia*). – Conservation Biology, **13** (5): 990–999.

KENNETT, R., D. A. FORDHAM, E. AIACS, B. COREY & A. GEORGES (2014): *Chelodina oblonga* GRAY 1841 – Northern Snake-Necked Turtle. – *In*: RHODIN, A. G. J., P. C. H. PRITCHARD, P. P. VAN DIJK, R. A. SAUMURE, K. A. BUHLMANN, J. B. IVERSON & R. A. MITTERMEIER (Eds.): Conservation Biology of Freshwater Turtles and Tortoises: A Compilation Project of the IUCN/SSC Tortoise and Freshwater Turtle Specialist Group. – Chelonian Research Monographs, Lunenburg, **5** (7): 077.1–077.13.

KUCHLING, G. (2010): Taxonomy and nomenclature of the longneck turtle (genus *Chelodina*) from south-western Australia. – Records of the Western Australian Museum, Perth, **25**: 449–454.

KWET, A. (2014): „Hoseritis" – taxonomischer Vandalismus in Australien. – Minor, Leipzig, **13** (2): 28–30.

LAURIN, M. & R. R. REISZ (1995): A reevaluation of early amniote phylogeny. – Zoological Journal of the Linnean Society, London, **113**: 165–223.

LE, M., B N. REID, W. P. MCCORD, E. NARO-MACIEL, C. J. RAXWORTHY, G. AMATO & A. GEORGES (2013): Resolving the phylogenetic history of the short-necked turtles, genera *Elseya* and *Myuchelys* (Testudines: Chelidae) from Australia and New Guinea. – Molecular Phylogenetics and Evolution, **68** (2): 251–258.

LEGLER, J. M. (1981): The taxonomy, distribution, and ecology of Australian freshwater turtles (Testudines: Pleurodira: Chelidae). – National Geographic Society Research Reports, Washington, **13**: 391–404.

LOURENÇO, J. M., J. CLAUDE, N. GALTIER & Y. CHIARI (2012): Dating cryptodiran nodes: origin and diversification of the turtle superfamily Testudinoidea. – Molecular Phylogenetics and Evolution, **62** (1): 496–507.

MCCORD, W. P. & M. JOSEPH-OUNI (2007a): A new species of *Chelodina* (Testudines: Chelidae) from southwestern New Guinea (Papua, Indonesia). – Reptilia (GB), Barcelona, **52**: 47–52.

MCCORD, W. P. & M. JOSEPH-OUNI (2007b): A new genus of Australian longneck turtle (Testudines: Chelidae) and a new species of *Macrochelodina* from the Kimberley region of Western Australia (Australia). – Reptilia (GB), Barcelona, **55**: 56–64.

MCCORD, W. P. & S. A. THOMSON (2002): A new species of *Chelodina* (Testudines: Pleurodira: Chelidae) from Northern Australia. – Journal of Herpetology, **36** (2): 255–267.

MCCORD, W. P., J. CANN & M. JOSEPH-OUNI (2003): A taxonomic assessment of *Emydura* (Testudines: Chelidae) with descriptions of new subspecies from Queensland, Australia. – Reptilia (GB), Castelldefels, **27**: 59–63.

MICHAEL, D. & D. G. LINDERMEYER (2010): Reptiles of the NSW Murray Catchment: A Guide to Their Identification, Ecology and Conservation. – Canberra (CSIRO Publishing), 248 pp.

MŁYNARSKI, M. (1976): Handbuch der Palaeoherpetologie. Teil 7. Testudines. – Stuttgart (Fischer Verlag), 130 pp.

MODESTO, S. P. & J. S. ANDERSON (2004): The phylogenetic definition of of Reptilia. – Systematic Biology, Oxford, **53** (5): 815–821.

MOLFETTI, E., S. TORRES VILAÇA, J. Y. GEORGES, V. PLOT, E. DELCROIX, R. LE SCAO, A. LAVERGNE, S. BARRIOZ, F. RODRIGUES DOS SANTOS & B. DE THOIY (2013): Recent demographic history and present fine-scale structure in the Northwest Atlantic Leatherback (*Dermochelys coriacea*) turtle population. – PLOS ONE 8:e58061.

NARO-MACIEL, E., M. LE, N. N. FITZSIMMONS & G. AMATO (2008): Evolutionary relationships of marine turtles: a molecular phylogeny based on nuclear and mitochondrial genes. – Molecular Phylogenetics and Evolution, **49** (2): 659–662.

NARO-MACIEL, E., B. N. REID, S. E. ALTER, G. AMOTO, K. A. BJORNDAL, A. B. BOLTEN, M. MARTIN, C. J. NAIRN, B. SHAMBLIN & O. PINEDA-CATALAN (2014): From refugia to rookeries: phylogeography of Atlantic green turtles. – Journal of Experimental Marine Biology and Ecology, **461**: 306–316.

OGILBY, J. B. (1890): Description of a new Australian tortoise. – Records of the Australian Museum, Sydney, **19** (1):56–59.

OKAMOTO, K. & N. KAMEZAKI (2014): Morphological variation in *Chelonia mydas* (LINNAEUS, 1758) from the coastal water of Japan, with special reference to the turtles allied to *Chelonia mydas agassizii* BOCOURT, 1868. – Current Herpetology, Kyoto, **33** (1): 46–56.

OKAYAMA, T., R. DÍAZ-FERNANDEZ, Y. BABA, M. HALIM, O. ABE, N. AZENO & H. KOIKE (1999): Genetic diversity of the hawksbill turtle in the Indo-Pacific and Caribbean regions. – Chelonian Conservation and Biology, Lunenburg, **3** (2):362–367.

PARHAM, J. F. & D. E. FASTOVSKY (1997): The phylogeny of cheloniid sea turtles revisited. – Chelonian Conservation and Biology, Lunenburg, **2** (4): 548–554.

PARHAM, J. F. & N. D. PYENSON (2010): New sea turtle from the Miocene of Peru and the iterative evolution of feeding ecomorphologies since the Cretaceous. – Journal of Paleontology, Boulder, **84** (2): 231–247.

PARHAM, J. F. & G. R. ZUG (1996): *Chelonia agassizii* – valid or not? – Marine Turtle Newsletter, **72**: 2–5.

PEREIRA, A. G., J. STERLI, F. R. R. MOREIRA & C. G. SCHRAGO (2017): Multilocus phylogeny and statistical biogeography clarify the evolutionary history of major lineages of turtles. – Molecular Phylogenetics and Evolution, **113**: 59–66.

PRITCHARD, P. C. H. (1984): Piscivory in turtles, and evolution of the long-necked Chelidae. – *In*: FERGUSON, M. W. (Ed.): The Structure, Development and Evolution of Reptiles. – Symposia of the Zoological Society of London, London, **52**: 87–110.

PRITCHARD, P. C. H. & P. TREBBAU (1984): The Turtles of Venezuela. – Society for the Study of Amphibians and Reptiles, Oxford, 403 pp

RHODIN, A. G. J. & V. R. GENORUPA (2000): Conservation status of freshwater turtles in Papua New Guinea. – Chelonian Research Monographs, Lunenburg, **2**: 129–136.

RHODIN, A. G. J. & R. A. MITTERMEIER (1976): *Chelodina parkeri*, a new species of chelid turtle from New Guinea, with a discussion of *Chelodina siebenrocki* WERNER, 1901. – Bulletin of the Museum of Comparative Zoology, Cambridge, **147** (11): 465–488.

RHODIN, A. G. J., P. P. VAN DIJK & J. F. PARHAM (2008): Turtles of the world: annotated checklist of taxonomy and synonymy. – *In*: RHODIN, A. G. J., P. C. H. PRITCHARD, P. P. VAN DIJK, R. A. SAUMURE, K. A. BUHLMANN & J. B. IVERSON (Eds.): Conservation Biology of Freshwater Turtles and Tortoises: A Compilation Project of the IUCN/SSC Tortoise and Freshwater Turtle Specialist Group. – Chelonian Research Monographs, Lunenburg, **5** (1): 000.1–000.38.

RHODIN, A. G. J., H. KAISER, P. P. VAN DIJK, W. WÜSTER, M. O'SHEA, M. ARCHER, M. AULIYA, L. BOITANI, R. BOUR, V. CLAUSNITZER, T. CONTRERAS-MACBEATH, B. I. CROTHER, J. M. DAZA, C. A. DRISCOLL, O. FLORES-VILLELA, J. FRAZIER, U. FRITZ, A. GARDNER, C. GASCON, A. GEORGES, F. GLAW, F. G. GRAZZIOTIN, C. P. GROVES, G. HASZPRUNAR, P. HAVAŠ, J. M. HERO, M. HOFFMANN, M. S. HOOGMOED, B. D. HORNE, J. B. IVERSON, M. JÄCH, C. L. JENKINS, R. K. B. JENKINS, A. R. KIESTER, J. S. KEOGH, T. E. LACHER JR., J. E. LOVICH, L. LUISELLI, D .L. MAHLER, D. MALLON, R. MAST, R. W. MCDIARMID, J. MEASEY, R. A. MITTERMEIER, S. MOLUR, V. MOSSBRUGGER, R. MURPHY, D. NAISH, M. NIEKISCH, H. OTA, J. F. PARHAM, M. J. PARR, N. J. PILCHER, R. H. PINE, A. B. RYLANDS, J. G. SANDERSON, J. SAVAGE, W. SCHLEIP, G. J. SCROCCHI, H. B. SHAFFER, E. N. SMITH, R. SPRACKLAND, S. N. STUART, H. VETTER, L. J. VITT, T. WALLER, G. WEBB, E. O. WILSON, H. ZAHER, AND S. THOMSON (2015): Comment on *Spracklandus* HOSER, 2009 (Reptilia, Serpentes, Elapidae): request for confirmation of the availability of the generic name and for the nomenclatural validation of the journal in which it was published. (Case 3601; see BZN 70: 234–237; 71: 30–38, 133–135, 181–182, 252–253). – Bulletin of Zoological Nomenclature, London, **72** (1): 65–78.

SAVAGE, J. M. (2007): Comment on the proposed precedence of *Chelodina rugosa* OGILBY, 1890 (currently *Macrochelodina rugosa*; Reptilia, Testudines) over *Chelodina oblonga* GRAY, 1841 (Case 3351). – Bulletin of Zoological Nomenclature, London, **63** (3): 187.

SEDDON, J., A. GEORGES, P. BAVERSTOCK & W. P. MCCORD (1997): Phylogenetic relationships of chelid turtles (Pleurodira: Chelidae) based on mitochondrial 12S rRNA gene sequence variation. – Molecular Phylogenetics and Evolution, **7** (1): 55–61.

SHAMBLIN, B. M., K. A BJORNDAL, A. B. BOLTEN, Z. M. HILLIS-STARR, I. LUNDGREN, E. NARO-MACIEL & C. J. NAIRN (2012): Mitogenomic sequences better resolve stock structure of southern Greater Caribbean green turtle rookeries. – Molecular Ecology, **21** (10): 2330–2340.

SHAMBLIN, B. M., A. B. BOLTEN, F. A. ABREU-GROBOIS, K. A. BJORNDAL, L. CARDONA, C. CARRERAS, M. CLUSA, C. MONZÓN-ARGÜELLO, C. J. NAIRN, J. T. NIELSEN, R. NEL, L. S. SOARES, K. R. STEWART, S. T. VILAÇA, O. TÜRKOZAN, C. YILMAZ & P. H. DUTTON (2014): Geographic patterns of genetic variation in a broadly distributed marine vertebrate: new insights into the Loggerhead Turtle stock structure from expanded mitochondrial DNA sequences. – PLoS One 9:e85956.

SMITH, H. M. & R. B. SMITH (1980): Synopsis of the Herpetofauna of Mexico. Vol.VI. Guide to Mexican Turtles. – North Bennington (Johnson), 367 pp.

SPINKS, P. Q., A. GEORGES & H. B. SHAFFER (2015): Phylogenetic Uncertainty and Taxonomic Re-revisions: An Example from the Australian Short-necked Turtles (Testudines: Chelidae). – Copeia, Lawrence, **103** (3): 536–540.

THOMSON, S. (2000): On the identification of the holotype of *Chelodina oblonga* (Testudinata: Chelidae) with a discussion of the taxonomic implications. – Chelonian Conservation and Biology, Lunenburg, **3** (4): 74–749.

THOMSON, S. A. (2006): ICZN Case 3351. *Chelodina rugosa* OGILBY, 1890 (currently *Macrochelodina rugosa*; Reptilia, Testudines): proposed precedence over *Chelodina oblonga* GRAY, 1841. – Bulletin of Zoological Nomenclature, London, **63** (3): 187–193.

THOMSON, S. A. (2007): Comment on the proposed precedence of *Chelodina rugosa* OGILBY, 1890 (currently *Macrochelodina rugosa*; Reptilia, Testudines) over *Chelodina oblonga* GRAY, 1841. – Bulletin of Zoological Nomenclature, London, **64**: 127–128.

THOMSON, S. & A. GEORGES (2009): *Myuchelys* gen. nov. – a new genus for *Elseya latisternum* and related forms of Australian freshwater turtle (Testudines: Pleurodira: Chelidae). – Zootaxa, Auckland, **2053**: 32–42.

THOMSON, S. & A. GEORGES (2016): A new species of freshwater turtle of the genus *Elseya* (Testudinata: Pleurodira: Chelidae) from the Northern Territory of Australia. – Zootaxa, Auckland, **4061** (1): 18–28.

THOMSON, S., R. KENNETT & A. GEORGES (2000): A new species of long-necked turtle (Testudines: Chelidae) from the Arnhem Land Plateau, Northern Territory, Australia. – Chelonian Conservation and Biology, Lunenburg, **3** (4): 675–685.

THOMSON, S. A., Y. AMEPOU, J. ANAMIATO & A. GEORGES (2015): A new species and subgenus of *Elseya* (Testudines: Pleurodira: Chelidae) from New Guinea. – Zootaxa, Auckland, **4006** (1): 059–082.

TKACH, V. V. & S. D. SNYDER (2008): *Aptorchis glandularis* n. sp. (Digenea: Plagiorchioidea) from the northwestern red-faced turtle, *Emydura australis*, (Pleurodira: Chelidae) in the Kimberley, Western Australia. – Journal of Parasitology, Lawrence, **94** (4): 918–924.

TODD, E. V., D. BLAIR, S. FARLEY, L. FARRINGTON, N. N. FITZSIMMONS, A. GEORGES, C. J. LIMPUS & D. R. JERRY (2013): Contemporary genetic structure reflects historical drainage isolation in an Australian snapping turtle, *Elseya albagula*. – Zoological Journal of the Linnean Society, London, **169** (1): 200–214.

TODD, E. V., D. BLAIR, A. GEORGES, V. LUKOSCHEK & D. R. JERRY (2014a): A biogeographical history and timeline for the evolution of Australian snapping turtles (*Elseya*: Chelidae) in Australia and New Guinea. – Journal of Biogeography, **41** (5): 905–918.

TODD, E. V., D. BLAIR & D. R. JERRY (2014b): Influence of drainage divides versus arid corridors on genetic structure and demography of a widespread freshwater turtle, *Emydura macquarii krefftii*, from Australia. – Ecology and Evolution, **4** (5): 606–622.

TUCKER, A. D., N. N. FITZSIMMONS & F. R. GOVEDICH (2005): Euhirudinea from Australian Turtles (*Chelodina burrungandjii* and *Emydura australis*) of the Kimberley Plateau, Western Australia, Australia. – Comparative Parasitology, Lawrence, **72** (2): 241–244.

TURTLE TAXONOMY WORKING GROUP [BICKHAM, J. W., J. B. IVERSON, J. F. PARHAM, H. D. PHILIPPEN, A. G. J. RHODIN, H. B. SHAFFER, P. Q. SPINKS & P. P. VAN DIJK] (2007): An Annotated List of Modern Turtle Terminal Taxa with Comments on Areas of Taxonomic Instability and Recent Change. – *In*: SHAFFER, H. B., N. N. FITZSIMMONS, A. GEORGES & A. G. J. RHODIN (Eds.): Defining Turtle Diversity: Proceedings of a Workshop on Genetics, Ethics, and Taxonomy of Freshwater Turtles and Tortoises. – Chelonian Research Monographs, Lunenburg, **4**: 173–199.

TURTLE TAXONOMY WORKING GROUP [RHODIN, A. G. J., J. F. PARHAM, P. P. VAN DIJK & J. B. IVERSON] (2009): Turtles of the world: an annotated checklist of taxonomy and synonymy, 2009 update, with conservation status summary. – *In*: RHODIN, A. G. J., P. C. H. PRITCHARD, P. P. VAN DIJK, R. A. SAUMURE, K. A. BUHLMANN & J. B. IVERSON (Eds.): Conservation Biology of Freshwater Turtles and Tortoises: A Compilation Project of the IUCN/SSC Tortoise and Freshwater Turtle Specialist Group. – Chelonian Research Monographs, Lunenburg, **5** (2): 000.39–000.84.

TURTLE TAXONOMY WORKING GROUP [RHODIN, A. G. J., P. P. VAN DIJK, J. B. IVERSON & H. B. SHAFFER] (2010): Turtles of the world, 2010 update: annotated checklist of taxonomy, synonymy, distribution, and conservation status. – *In*: RHODIN, A. G. J., P. C. H. PRITCHARD, P. P. VAN DIJK, R. A. SAUMURE, K. A. BUHLMANN, J. B. IVERSON & R. A. MITTERMEIER (Eds.): Conservation Biology of Freshwater Turtles and Tortoises: A Compilation Project of the IUCN/SSC Tortoise and Freshwater Turtle Specialist Group. – Chelonian Research Monographs, Lunenburg, **5** (3): 000.85–000.164.

TURTLE TAXONOMY WORKING GROUP [VAN DIJK, P. P., J. B. IVERSON, H. B. SHAFFER, R. BOUR & A. G. J. RHODIN] (2011): Turtles of the world, 2011 update: annotated checklist of taxonomy, synonymy, distribution, and conservation status. – *In*: RHODIN, A. G. J., P. C. H. PRITCHARD, P. P. VAN DIJK, R. A. SAUMURE, K. A. BUHLMANN, J. B. IVERSON & R. A. MITTERMEIER (Eds.): Conservation Biology of Freshwater Turtles and Tortoises: A Compilation Project of the IUCN/SSC Tortoise and Freshwater Turtle Specialist Group. – Chelonian Research Monographs, Lunenburg, **5** (4): 000.165–000.242.

TURTLE TAXONOMY WORKING GROUP [VAN DIJK, P. P., J. B. IVERSON, H. B. SHAFFER, R. BOUR & A. G. J. RHODIN] (2012): Turtles of the world, 2012 update: annotated checklist of taxonomy, synonymy, distribution, and conservation status. – *In*: RHODIN, A. G. J., P. C. H. PRITCHARD, P. P. VAN DIJK, R. A. SAUMURE, K. A. BUHLMANN, J. B. IVERSON & R. A. MITTERMEIER (Eds.): Conservation Biology of Freshwater Turtles and Tortoises: A Compilation Project of the IUCN/SSC Tortoise and Freshwater Turtle Specialist Group. – Chelonian Research Monographs, Lunenburg, **5** (5): 000.243–000.328.

TURTLE TAXONOMY WORKING GROUP [VAN DIJK, P. P., J. B. IVERSON, A. G. J. RHODIN, H. B. SHAFFER & R. BOUR] (2014): Turtles of the world, 7th edition: annotated checklist of taxonomy, synonymy, distribution with maps, and conservation status. – *In*: RHODIN, A. G. J., P. C. H. PRITCHARD, P. P. VAN DIJK, R. A. SAUMURE, K. A. BUHLMANN, J. B. IVERSON & R. A. MITTERMEIER (Eds.): Conservation Biology of Freshwater Turtles and Tortoises: A Compilation Project of the IUCN/SSC Tortoise and Freshwater Turtle Specialist Group. – Chelonian Research Monographs, Lunenburg, **5** (7): 000.329–000.479.

TURTLE TAXONOMY WORKING GROUP [RHODIN, A. G. J., J. B. IVERSON, R. BOUR, U. FRITZ, A. GEORGES, H. B. SHAFFER & P. P. VAN DIJK] (2017): Turtles of the world: annotated checklist and atlas of taxonomy, synonymy, distribution, and conservation status (8th Ed.). – *In*: RHODIN, A. G. J., J. B. IVERSON, P. P. VAN DIJK, R. A. SAUMURE, K. A. BUHLMANN, P. C. H. PRITCHARD & R. A. MITTERMEIER (Eds.): Conservation Biology of Freshwater Turtles and Tortoises: A Compilation Project of the IUCN/SSC Tortoise and Freshwater Turtle Specialist Group. – Chelonian Research Monographs, Lunenburg, **7**, 292 pp. doi: 10.3854/crm.7.checklist.atlas.v8.2017.

VETTER, H. (2014): „Hoseritis", 2. Teil. – Minor, Leipzig, **13** (3): 17–18.

WELLS, R. W. (2002a): A New Subspecies of *Carettochelys* (Reptilia: Carettochelydidae) from Northern Australia – *Carettochelys insculpta canni* ssp. nov. – Australian Biodiversity Record, Cowra, **2002** (1): 1–7.

WELLS, R. W. (2002b): Taxonomic notes on some Australian freshwater turtles of the genera *Chelodina* and *Elseya*. – Australian Biodiversity Record, Cowra, **2002** (2): 1–30.

WELLS, R. W. (2007a): Some taxonomic and nomenclatural considerations on the class Reptilia in Australia. Notes on the recently described freshwater turtle *Chelodina canni* MCCORD and THOMSON, 2002 and a redescription of *Chelodina rankini* WELLS and WELLINGTON, 1985. – Australian Biodiversity Record, Cowra, **2007** (1): 1–5.

WELLS, R. W. (2007b) Some taxonomic and nomenclatural considerations on the class Reptilia in Australia. Some comments on the *Elseya dentata* (GRAY, 1863) complex with redescriptions of the Johnstone River Snapping Turtle, *Elseya stirlingi* WELLS and WELLINGTON, 1985 and the Alligator River Snapping Turtle, *Elseya jukesi* WELLS 2002. – Australian Biodiversity Record, Cowra, **2007** (2): 1–12.

WELLS, R. W. (2007c): Some taxonomic and nomenclatural considerations on the Class Reptilia in Australia. A new genus of the Family Chelidae from eastern Australia. – Australian Biodiversity Record, Cowra, **2007** (3): 1–13.

WELLS, R. W. (2009): Some taxonomic and nomenclatural considerations on the Class Reptilia in Australia. A new species of freshwater turtle in the genus *Wollumbinia* WELLS 2007 (Reptilia: Chelidae) from eastern Australia. – Australian Biodiversity Record, Cowra, **2009** (1): 1–12.

WELLS, R. W. & C. R. WELLINGTON (1985): A classification of the Amphibia and Reptilia of Australia. – Australian Journal of Herpetology, Supplementary Series, Sydney, **1**: 1–61.

WHITE, A. W., T. H. WORTHY, S. HAWKINS, S. BEDFORD & M. SPRIGGS (2010): Megafaunal meiolaniid horned turtles survived until early human settlement in Vanuatu, Southwest Pacific. – Proceedings of the National Academy of Sciences of the United States of America, Washington, **107** (35): 15512–15516.

WILSON, S. & G. SWAN (2008): A Complete Guide to Reptiles of Australia. 2nd edition. – Sydney, Auckland, London & Cape Town (Reed New Holland), 512 pp.

WILSON, S. & G. SWAN (2013) A Complete Guide to Reptiles of Australia. 4th revised edition. – London, Sydney, Cape Town & Auckland (New Holland Publishers), 592 pp.

WORRELL, E. (1970): Reptiles of Australia. – Sydney (Angus and Robertson), 169 pp.

ZANGERL, R. (1958): Die oligozänen Meerschildkröten von Glarus. – Schweizerische Paläontologische Abhandlungen, Basel, **73**: 1–56.

ZANGERL, R. & W. D. TURNBULL (1955): *Procolpochelys grandaeva* (LEIDY), an early carettine sea turtle. – Fieldiana Zoology, Chicago, **37** (12) 345–382.

ZHANG, X., P. J. UNMACK, G. KUCHLING, Y. WANG & A. GEORGES (2017): Resolution of the enigmatic phylogenetic relationships of the critically endangered Western Swamp Tortoise *Pseudemydura umbrina* (Pleurodira: Chelidae) using a complete mitochondrial genome. – Molecular Phylogenetics and Evolution, **115**: 58–61.

ZHUO, W., J. PASCUAL-ANAYA, A. ZADISSA, W. LI, Y. NIIMURA, Z. HUANG, C. LI, S. WHITE, Z. XIONG, D. FANG, B. WANG, Y. MING, Y. CHEN, Y. ZHENG, S. KURAKU, M. PIGNATELLI, J. HERRERO, K. BEAL, M. NOZAWA, Q. LI, J. WANG, H. ZHANG, L. YU, S. SHIGENOBU, J. WANG, J. LIU, P. F., S. SEARLE, J. WANG, S. KURATANI, Y. YIN, B. AKEN, G. ZHANG & N. IRIE (2013): The draft genomes of softshell turtle and green sea turtle yield insights into the development and evolution of the turtlespecific body plan. – Nature Genetics, London, **45** (6): 701–706.

Acknowledgments/Danksagung

Pictures taken in the wild were principally preferred when selecting the material for this volume. However, in some cases captive animals are shown. Three taxa are only represented by photos of museum specimens/*Bei der Auswahl der Bilder wurde Aufnahmen im natürlichen Lebensraum der Vorzug gegeben. In Einzelfällen musste auf Fotos von Tieren in menschlicher Obhut zurückgegriffen werden. Von drei Arten lagen nur Bilder von Museumsexemplaren vor.*

For providing us with their pictures we are grateful to the following persons/*Den folgenden Personen möchten wir hiermit für die Bereitstellung ihres Bildmaterials herzlich danken:*

Hans & Judy BESTE (Australia), David BETTINI (Australia), Eva BOOGAARD (Australia), Gavin BROWN (Australia), John CANN (Australia), Rohan CLARKE (Australia), Adam ELLIOTT (Australia), Damian GOODALL (Australia), Gregory COSENTINO (USA), Cris HAGEN (USA), James H. HARDING (USA), David HEGNER (Czech Republic), Alex HOWARD (Australia), Gerald KUCHLING (Australia), Jiri LOCHMAN (Australia), Craig LATTA (Australia), Jérôme MARAN (France), J. Lindley MCKAY (Australia), Rune MIDTGAARD (Denmark), Mark O'SHEA (United Kingdom), Hynek PROKOP (Czech Republic), Jens REISSIG (South Africa), Anders RHODIN (USA), Gunther SCHMIDA (Australia), Ruchira SOMAWEERA (Sri Lanka/Australia), Alex STEFFE (Australia), Steffen SZYMANSKI (Germany), Peter Paul VAN DIJK (USA), Steve WILSON (Australia), Belinda WRIGHT (Australia).

Contributors to the Appendix/
Fotografen, die uns beim Anhang unterstützt haben:
Torsten BLANCK (Austria), Vinicus Carvalho (Brazil), Carlos del Valle (Colombia), CHAN TAI MAN (Hong Kong, P.R. China), Tomas DIAGNE (Senegal), Maren GAULKE (Germany), Brian D. HORNE (USA), Flora IHLOW (Germany), Marco Antonio LOPEZ-LUNA (Mexico), Tomáš MAZUCH (Czech Republic), Enrico MORAWA (Germany), Peter PRASCHAG (Austria), Peter A. SCOTT (USA), Pavel ŠIROKY (Czech Republic), Robert T. ZAPPALORTI (USA), Richard C. VOGT (Brazil).

We would not have been able to gather this collection of photos without the help of the following friends and colleagues. We would like to express our gratitude for their support in particular to/
Ohne die engagierte Hilfe der folgenden Freunde und Kollegen wäre es nicht möglich gewesen, die Abbildungen für den vorliegenden Band zusammenzustellen. Herzlich danken wir:
Bob & Sheri ASHLEY, John CANN, Patrick DAVID, Bernd EIDENMÜLLER, Balázs Farkas, Uwe FRITZ, Cris HAGEN, Andreas S. HENNIG, Gerald KUCHLING, William P. MCCORD, Anders RHODIN, Ross SADLIER, Gunther SCHMIDA, Mario VARGAS RAMIREZ, Steve WILSON.

RT05611-4 *Caretta caretta* (LINNAEUS, 1758)
Ningaloo Marine Park, WA, Australia
135–213 (?) cm

photo: E. BOOGAARD/LOCHMAN LT

RT05612-3 *Caretta caretta* (LINNAEUS, 1758)
135–213 (?) cm
Subadult

photo: R. SOMAWEERA

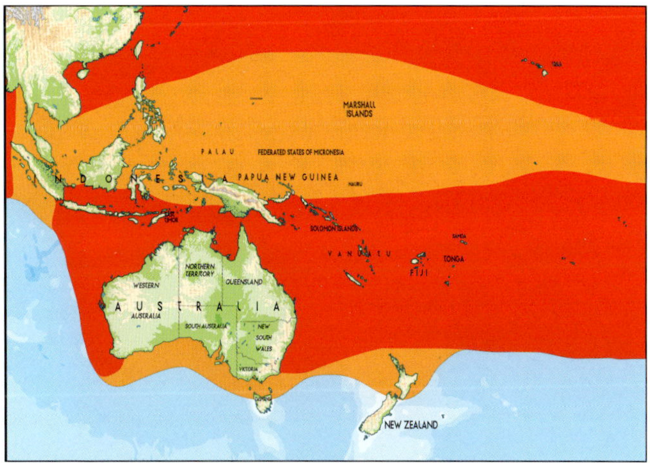

Caretta caretta (red: nesting and feeding; orange: vagrant)

RT05613-4 *Caretta caretta* (LINNAEUS, 1758)
Mon Repos, QLD, Australia
135–213 (?) cm

photo: S. WILSON

RT05614-3 *Caretta caretta* (LINNAEUS, 1758)
135–213 (?) cm
Subadult

photo: R. SOMAWEERA

RT05615-4 *Caretta caretta* (LINNAEUS, 1758)
Dirk Hartog Island, Shark Bay, WA, Australia
135–213 (?) cm
Oviposition

photo: J. LOCHMAN/LOCHMAN LT

RT05616-1 *Caretta caretta* (LINNAEUS, 1758)
Mon Repos, QLD, Australia
135–213 (?) cm
Hatchling

photo: S. WILSON

RT05617-1 *Caretta caretta* (LINNAEUS, 1758)
Dirk Hartog Island, Shark Bay, WA, Australia
135–213 (?) cm
Hatchling

photo: J. LOCHMAN/LOCHMAN LT

RT05618-4 *Caretta caretta* (LINNAEUS, 1758)
Dirk Hartog Island, Shark Bay, WA, Australia
135–213 (?) cm
Oviposition
photo: J. LOCHMAN/LOCHMAN LT

RT05619 *Caretta caretta* (LINNAEUS, 1758)
135–213 (?) cm
Golden Bandicoot preying on freshly laid eggs, Barrow Island, WA, Australia
photo: J. LOCHMAN/LOCHMAN LT

RT05620 *Caretta caretta* (LINNAEUS, 1758)
135–213 (?) cm
Ghost Crab preying on a hatchling, Dirk Hartog Island, Shark Bay, WA, Australia
photo: J. LOCHMAN/LOCHMAN LT

RT05631-4 *Eretmochelys imbricata* (LINNÉ, 1766)
Bare Sand Island, NT, Australia
100–114 cm

photo: J. LOCHMAN/LOCHMAN LT

RT05632-4 *Eretmochelys imbricata* (LINNÉ, 1766)
Nagada Harbour, Madang Prov., Papua New Guinea
100–114 cm

photo: M. O'SHEA

RT05633-4 *Eretmochelys imbricata* (LINNÉ, 1766)
Bare Sand Island, NT, Australia
100–114 cm

photo: J. LOCHMAN/LOCHMAN LT

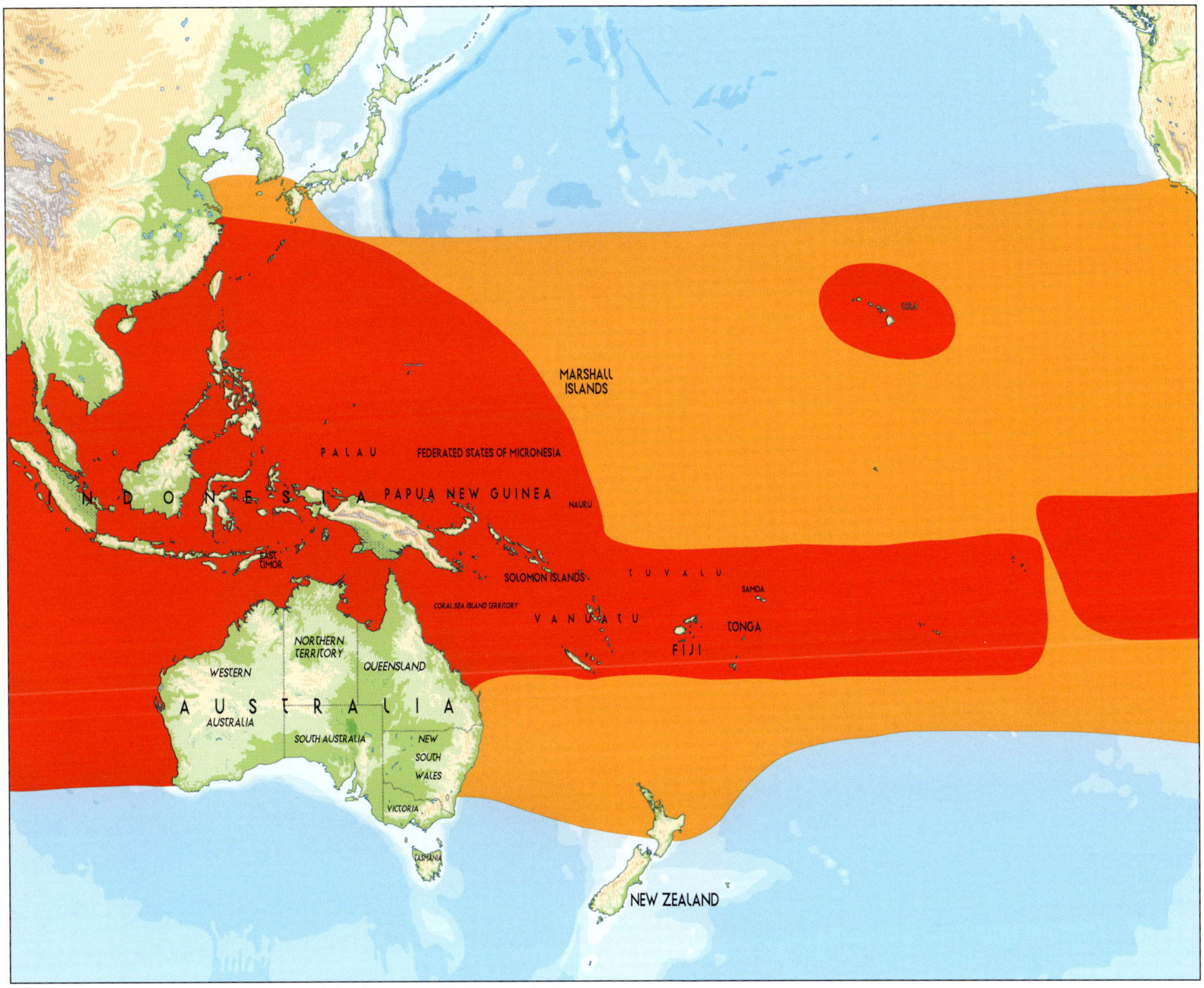

Eretmochelys imbricata (red: nesting and feeding; orange: vagrant)

RT05634-4 *Eretmochelys imbricata* (LINNÉ, 1766)

100–114 cm

photo: R. SOMAWEERA

RT05635-4 *Eretmochelys imbricata* (LINNÉ, 1766)

100–114 cm

photo: R. SOMAWEERA

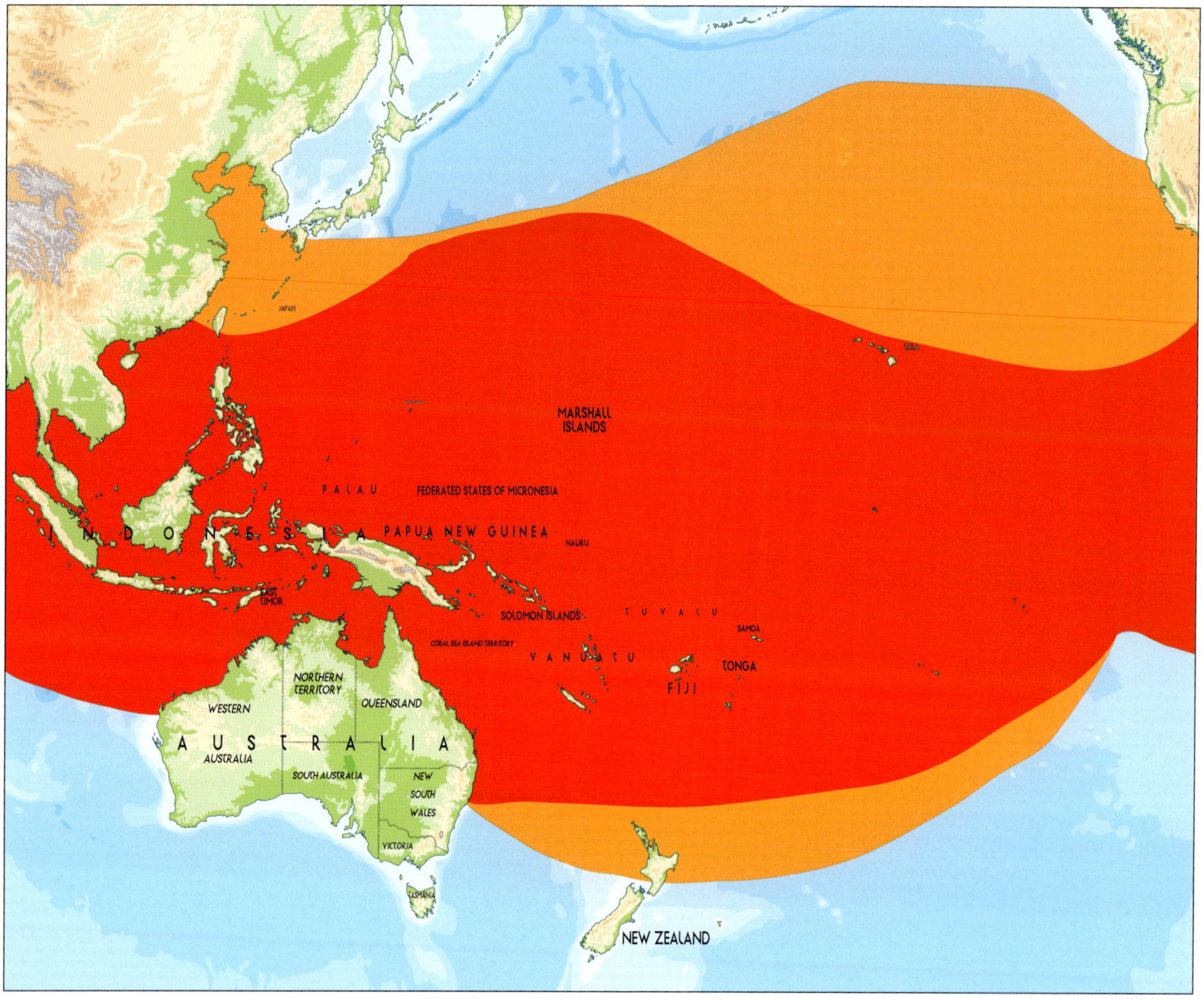

Lepidochelys olivacea (red: nesting and feeding; orange: vagrant)

RT05641-4 *Lepidochelys olivacea* (Eschscholtz, 1829)
Thailand
70–80 cm

photo: P.P. van Dijk

RT05642-1 *Lepidochelys olivacea* (Eschscholtz, 1829)

70–80 cm
Hatchling

photo: R. Somaweera

RT05651-4 *Chelonia mydas* (LINNAEUS, 1758)
Great Barrier Reef, QLD, Australia
140–153 cm

photo: A. STEFFE/LOCHMAN LT

RT05652-4 *Chelonia mydas* (LINNAEUS, 1758)
Barrow Island, WA, Australia
140–153 cm
Mating

photo: G. KUCHLING

RT05653-4 *Chelonia mydas* (LINNAEUS, 1758)
Barrow Island, WA, Australia
140–153 cm

photo: G. KUCHLING

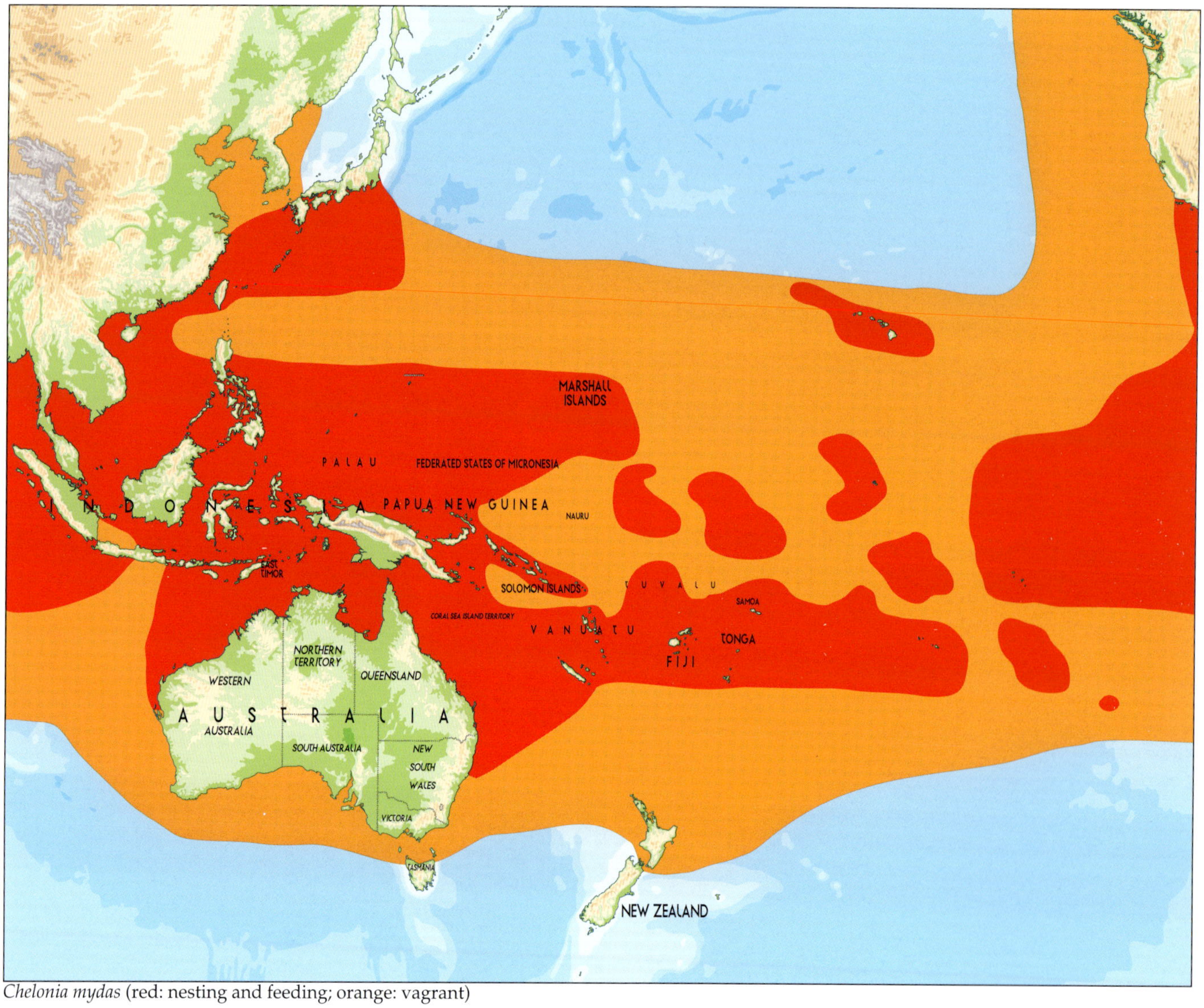

Chelonia mydas (red: nesting and feeding; orange: vagrant)

RT05654-4 *Chelonia mydas* (Linnaeus, 1758)
Bare Sand Island, NT, Australia
140–153 cm

photo: J. Lochman/Lochman LT

RT05655-4 *Chelonia mydas* (Linnaeus, 1758)
Hawaii, USA
140–153 cm

photo: C. Hagen

RT05656-4 *Chelonia mydas* (Linnaeus, 1758)
Barrow Island, WA, Australia
140–153 cm

photo: J. Lochman/Lochman LT

RT05657-4 *Chelonia mydas* (Linnaeus, 1758)
Barrow Island, WA, Australia
140–153 cm

photo: G. Kuchling

RT05658-3 *Chelonia mydas* (Linnaeus, 1758)

140–153 cm
Subadult

photo: R. Somaweera

RT05659-4 *Chelonia mydas* (LINNAEUS, 1758)
Barrow Island, WA, Australia
140–153 cm

photo: G. KUCHLING

RT05660-1 *Chelonia mydas* (LINNAEUS, 1758)

140–153 cm
Hatchling

photo: R. SOMAWEERA

RT05661-1 *Chelonia mydas* (LINNAEUS, 1758)
Barrow Island, WA, Australia
140–153 cm
Hatchlings

photo: G. KUCHLING

RT05671-4 *Natator depressus* (Garman, 1880)
Bare Sand Island, NT, Australia
110–115 cm

photo: J. Lochman/Lochman LT

RT05672-4 *Natator depressus* (Garman, 1880)
Mundabulangana Station, Pilbara, WA, Australia
110–115 cm

photo: G. Kuchling

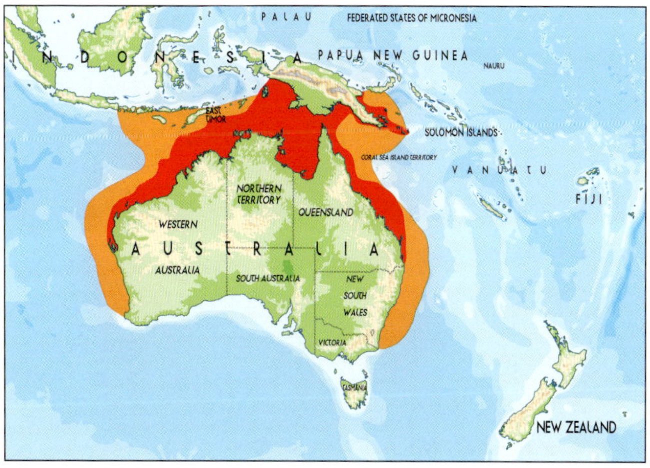

Natator depressus (red: nesting and feeding; orange: vagrant)

RT05673-4 *Natator depressus* (GARMAN, 1880)
Bare Sand Island, NT, Australia
110–115 cm
Oviposition
photo: J. LOCHMAN/LOCHMAN LT

RT05674-4 *Natator depressus* (GARMAN, 1880)
Bare Sand Island, NT, Australia
110–115 cm
Oviposition
photo: G. KUCHLING

RT05675 *Natator depressus* (GARMAN, 1880)
Bare Sand Island, NT, Australia
Tracks of a female returning to the sea after nesting
photo: J. LOCHMAN/LOCHMAN LT

RT05676-4 *Natator depressus* (Garman, 1880)
Bare Sand Island, NT, Australia
110–115 cm
photo: J. Lochman/Lochman LT

RT05677-1 *Natator depressus* (Garman, 1880)
Bare Sand Island, NT, Australia
110–115 cm
Hatching
photo: J. Lochman/Lochman LT

RT05678-4 *Natator depressus* (Garman, 1880)
Bare Sand Island, NT, Australia
110–115 cm
Female returning to the sea
photo: J. Lochman/Lochman LT

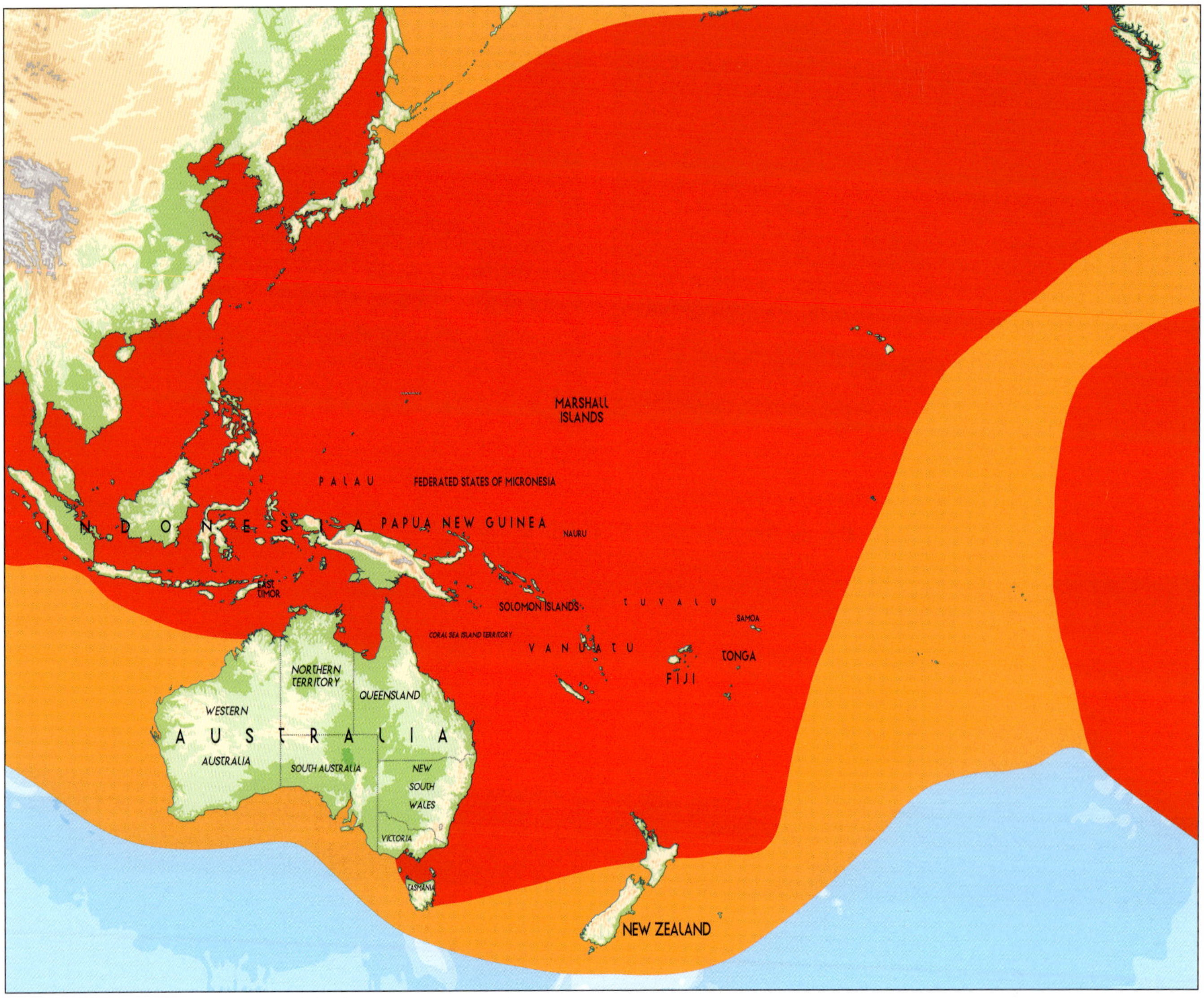

Dermochelys coriacea (red: nesting and feeding; orange: vagrant)

RT05681-1 *Dermochelys coriacea* (VANDELLI, 1761)

200–220 cm
Hatching

photo: M. O'SHEA

RT05682-1 *Dermochelys coriacea* (VANDELLI, 1761)

200–220 cm
Hatchling

photo: M. O'SHEA

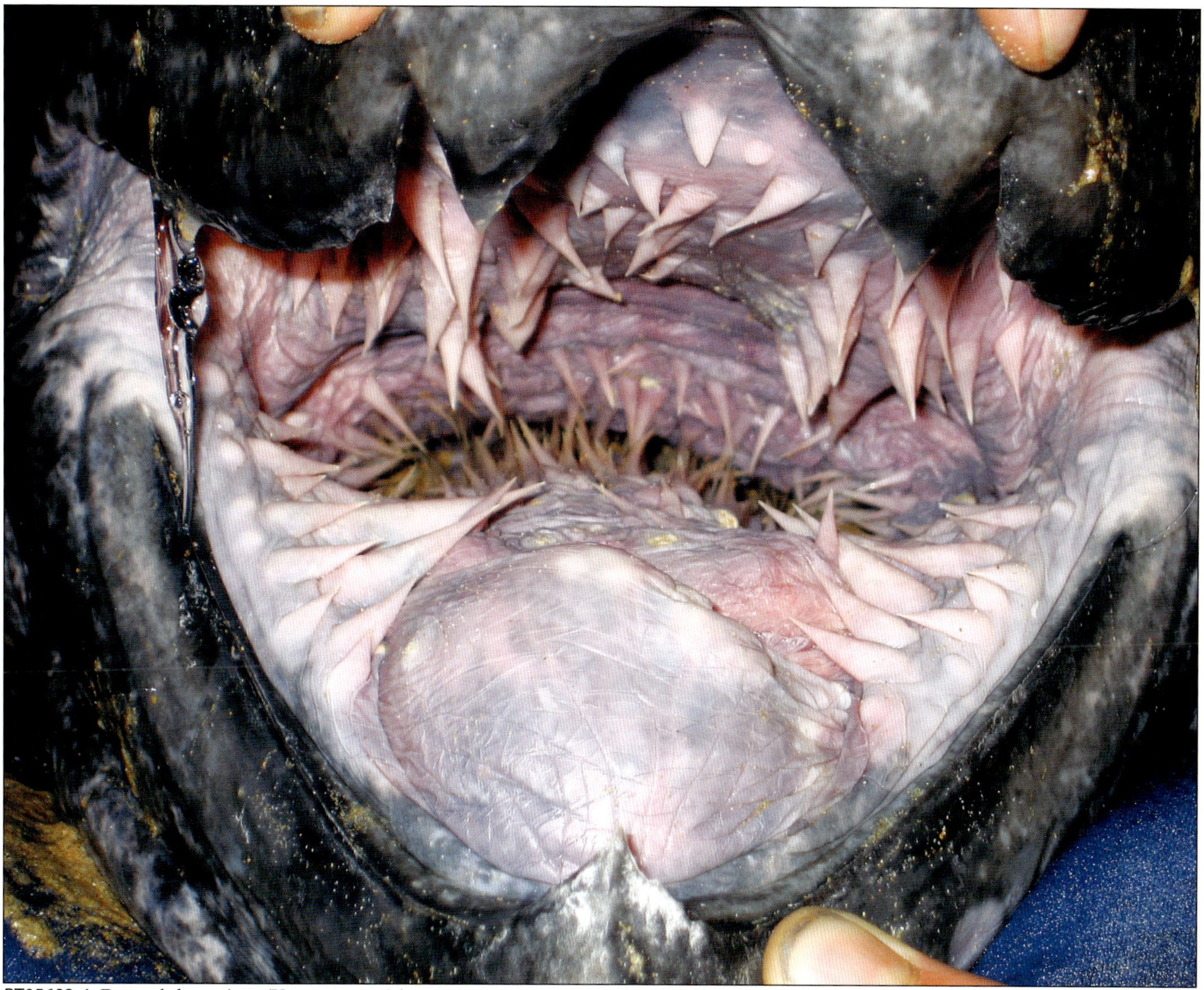

RT05683-4 *Dermochelys coriacea* (VANDELLI, 1761)

200–220 cm

photo: A.HOWARD courtesy of D. GOODALL

RT05684-4 *Dermochelys coriacea* (VANDELLI, 1761)

200–220 cm

photo: D. GOODALL

RT05685-4 *Dermochelys coriacea* (VANDELLI, 1761)
KwaZulu-Natal Prov., South Africa
200–220 cm
♀

photo: J. REISSIG

RT05691-3 *Carettochelys insculpta* RAMSAY, 1886
Southern Papua New Guinea
50–60 cm
Subadult

photo: M. O'SHEA

RT05692-4 *Carettochelys insculpta* RAMSAY, 1886
Daly River, NT, Australia
50–60 cm

photo: G. SCHMIDA

RT05693-4 *Carettochelys insculpta* RAMSAY, 1886
Daly River, NT, Australia
50–60 cm

photo: J. CANN

Carettochelys insculpta (red; ? – questionable occurrence; * – introduced populations)

RT05694-4 *Carettochelys insculpta* RAMSAY, 1886
Daly River, NT, Australia
50–60 cm

photo: G. SCHMIDA

RT05695-4 *Carettochelys insculpta* RAMSAY, 1886
Daly River, NT, Australia
50–60 cm

photo: G. KUCHLING

RT05696-5 *Carettochelys insculpta* RAMSAY, 1886
Daly River, NT, Australia
50–60 cm

photo: S. WILSON

RT05697-3 *Carettochelys insculpta* RAMSAY, 1886
Captive specimen
50–60 cm
Subadult

photo: J. MARAN

RT05698-2 *Carettochelys insculpta* Ramsay, 1886
Daly River, NT, Australia
50–60 cm
Juvenile

photo: G. Schmida

m ≈ ≅ ≣ ≗ 35°C ☺ 1 STOP

RT05699-5 *Carettochelys insculpta* Ramsay, 1886
Daly River, NT, Australia
50–60 cm

photo: S. Wilson

m ≈ ≅ ≣ ≗ 35°C ☺ 1 STOP

RT05701-4 *Carettochelys insculpta* Ramsay, 1886
Daly River, NT, Australia
50–60 cm

photo: G. Schmida

RT05702-2 *Carettochelys insculpta* Ramsay, 1886
Captive specimen
50–60 cm
Juvenile

photo: J. Lochman/Lochman LT

RT05703-2 *Carettochelys insculpta* Ramsay, 1886
Daly River, NT, Australia
50–60 cm
Juvenile

photo: G. Schmida

Pelochelys bibroni (red; ? – questionable occurrence),
Pelochelys signifera (blue)

RT05711-4 *Pelochelys bibroni* (Owen, 1853)
Lake Murray, Western Prov., Papua New Guinea
95–102 cm

photo: A. Rhodin

RT05721-4 *Pelochelys signifera* Webb, 2003
Maprik, East Sepik Prov., Papua New Guinea
95–102 cm

photo: A. Rhodin

RT05731-4 *Chelodina* (*Chelodina*) *canni* McCord & Thomson, 2002
Roper River, NT, Australia
17–27 cm

photo: G. Schmida

35°C 3

RT05732-4 *Chelodina* (*Chelodina*) *canni* McCord & Thomson, 2002
Roper River, NT, Australia
17–27 cm

photo: G. Schmida

35°C 3

RT05733-4 *Chelodina* (*Chelodina*) *canni* McCord & Thomson, 2002
Roper River, NT, Australia
17–27 cm

photo: G. Schmida

35°C 3

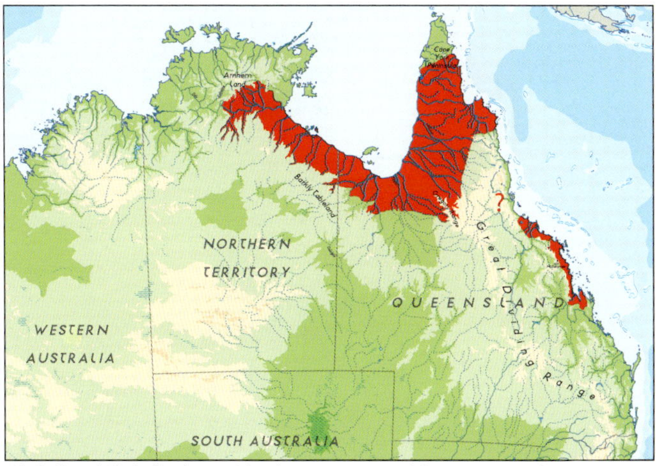

Chelodina (Chelodina) canni (red; ? – questionable occurrence)

RT05734-4 *Chelodina* (*Chelodina*) *canni* McCord & Thomson, 2002
Bowen, QLD, Australia
17–27 cm

photo: J. Cann

RT05735-4 *Chelodina* (*Chelodina*) *canni* McCord & Thomson, 2002
Proserpine, QLD, Australia
17–27 cm

photo: A. Elliott

RT057361-4 *Chelodina* (*Chelodina*) *canni* McCord & Thomson, 2002
Ross River area, QLD, Australia
17–27 cm

photo: S. Wilson

RT05737-4 *Chelodina* (*Chelodina*) *canni* McCord & Thomson, 2002
Muller's Lagoon, Bowen, QLD, Australia
17–27 cm

photo: G. Schmida

RT05741-4 *Chelodina (Chelodina) gunaleni* McCord & Joseph-Ouni, 2007
Uta River area, Papua, Indonesia
16–24 cm

photo: H. Prokop

35°C 3

RT05743-4 *Chelodina (Chelodina) gunaleni* McCord & Joseph-Ouni, 2007
Uta River area, Papua, Indonesia
16–24 cm
♀ + ♂

photo: H. Prokop

35°C 3

RT05743-4 *Chelodina (Chelodina) gunaleni* McCord & Joseph-Ouni, 2007
Captive specimen
16–24 cm
♀

photo: S. Szymanski

35°C 3

RT05744-1 *Chelodina (Chelodina) gunaleni* McCord & Joseph-Ouni, 2007
16–24 cm
Hatching

photo: S. Szymanski

35°C 3

RT05745-4 *Chelodina (Chelodina) gunaleni* McCord & Joseph-Ouni, 2007
Captive specimens
16–24 cm

photo: S. Szymanski

35°C 3

Chelodina (Chelodina) gunaleni (red),
Chelodina (Chelodina) novaeguineae (blue)

RT05751-4 *Chelodina (Chelodina) longicollis* (SHAW, 1794)
Condamine River catchment, QLD, Australia
25–29 cm

photo: G. SCHMIDA

35°C 3

RT05752-4 *Chelodina (Chelodina) longicollis* (SHAW, 1794)
Nerang River, NSW, Australia
25–29 cm

photo: G. SCHMIDA

35°C 3

RT05753-4 *Chelodina (Chelodina) longicollis* (Shaw, 1794)
Southeast QLD, Australia
25–29 cm

photo: G. Schmida

RT05754-4 *Chelodina (Chelodina) longicollis* (Shaw, 1794)
Condamine River catchment, QLD, Australia
25–29 cm

photo: G. Schmida

RT05755-4 *Chelodina (Chelodina) longicollis* (Shaw, 1794)
Southeast QLD, Australia
25–29 cm

photo: G. Schmida

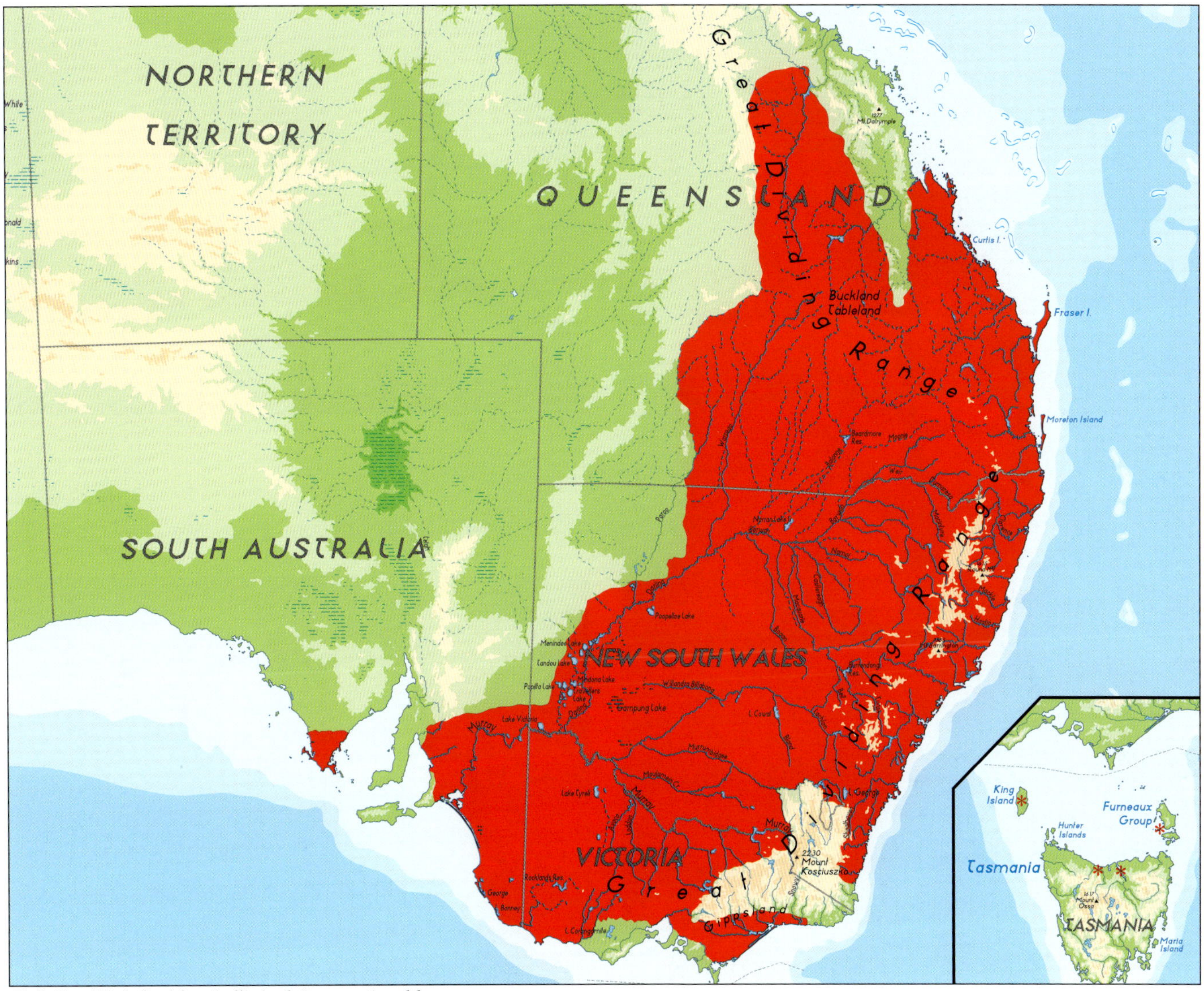

Chelodina (Chelodina) longicollis (red; ? – questionable occurrence; * – introduced populations)

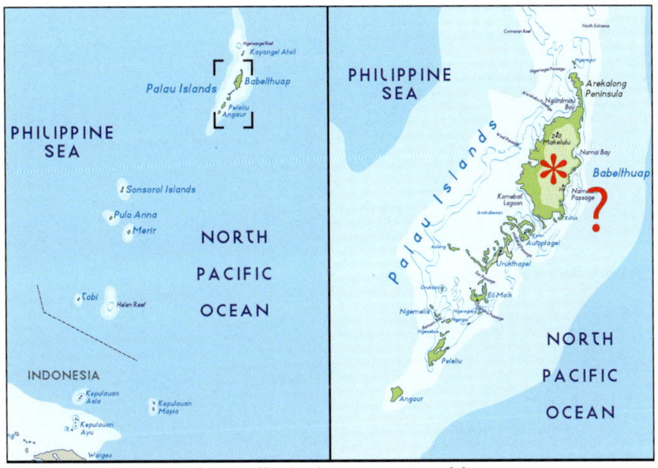

Chelodina (Chelodina) longicollis (red; ? – questionable occurrence; * – introduced populations)

Chelodina (Chelodina) longicollis (red; ? – questionable occurrence; * – introduced populations)

RT05756-4 *Chelodina* (*Chelodina*) *longicollis* (Shaw, 1794)
Sydney, NSW, Australia
25–29 cm

photo: G. Schmida

m ≋ ≅ ≡ 35°C ☹ 3

RT05757-4 *Chelodina* (*Chelodina*) *longicollis* (Shaw, 1794)
Captive specimens
25–29 cm
♀ + ♂

photo: H. Prokop

m ≋ ≅ ≡ 35°C ☹ 3

RT05758-4 *Chelodina* (*Chelodina*) *longicollis* (Shaw, 1794)
Namoi River drainage, Western NSW, Australia
25–29 cm
♂ + ♀

photo: G. Schmida

m ≋ ≅ ≡ 35°C ☹ 3

RT05759-2 *Chelodina* (*Chelodina*) *longicollis* (Shaw, 1794)
Captive-bred, QLD, Australia
25–29 cm
Juvenile

photo: G. Schmida

m ≋ ≅ ≡ 35°C ☹ 3

RT05760-2 *Chelodina* (*Chelodina*) *longicollis* (Shaw, 1794)
Captive-bred, QLD, Australia
25–29 cm
Juvenile

photo: G. Schmida

m ≋ ≅ ≡ 35°C ☹ 3

RT05761-4 *Chelodina (Chelodina) longicollis* (SHAW, 1794)
Sydney, NSW, Australia
25–29 cm

photo: G. SCHMIDA

35°C 3

RT05762-4 *Chelodina (Chelodina) longicollis* (SHAW, 1794)
Severn River, NSW, Australia
25–29 cm

photo: G. SCHMIDA

35°C 3

RT05763-4 *Chelodina (Chelodina) longicollis* (SHAW, 1794)
QLD, Australia
25–29 cm

photo: S. WILSON

35°C 3

RT05764-4 *Chelodina (Chelodina) longicollis* (SHAW, 1794)
Captive specimen
25–29 cm

photo: R. MIDTGAARD

35°C 3

RT05765-4 *Chelodina (Chelodina) longicollis* (SHAW, 1794)
Condamine River, NSW, Australia
25–29 cm

photo: G. SCHMIDA

35°C 3

RT05766-4 *Chelodina (Chelodina) longicollis* (SHAW, 1794)

25–29 cm

photo: R. SOMAWEERA

35°C 3

RT05767-4 *Chelodina* (*Chelodina*) *longicollis* (Shaw, 1794)
Colo River, Blue Mountains, NSW, Australia
25–29 cm

photo: G. Schmida

RT05768-4 *Chelodina* (*Chelodina*) *longicollis* (Shaw, 1794)
Erringibba NP, QLD, Australia
25–29 cm

photo: S. Wilson

RT0XXXX-1 *Chelodina* (*Chelodina*) *longicollis* (Shaw, 1794)
Habitat, QLD, Australia

photo: A. Elliott

RT05769-4 *Chelodina (Chelodina) longicollis* (Shaw, 1794)
Bald Rock Creek, NSW, Australia
25–29 cm
photo: A. Elliott

RT05771-4 *Chelodina (Chelodina) longicollis* (Shaw, 1794)
Biloela, QLD, Australia
25–29 cm
photo: A. Elliott

RT05772-1 *Chelodina (Chelodina) longicollis* (Shaw, 1794)
Captive specimen
25–29 cm
Hatchling
photo: A. Elliott

RT05773-2 *Chelodina (Chelodina) longicollis* (Shaw, 1794)
Captive specimen
25–29 cm
Juvenile
photo: A. Elliott

RT05774-4 *Chelodina (Chelodina) longicollis* (SHAW, 1794)
Gwydir River, NSW, Australia
25–29 cm

photo: A. ELLIOTT

35°C 3

RT05775-4 *Chelodina (Chelodina) longicollis* (SHAW, 1794)
Swan Hill, VIC, Australia
25–29 cm

photo: R. CLARKE

35°C 3

RT05781-4 *Chelodina (Chelodina) novaeguineae* BOULENGER, 1888
Merauke Regency, Papua, Indonesia
25–30 cm

photo: C. HAGEN

RT05782-4 *Chelodina (Chelodina) novaeguineae* BOULENGER, 1888
Captive specimen
25–30 cm

photo: G. COSENTINO

RT05783-4 *Chelodina (Chelodina) novaeguineae* BOULENGER, 1888
Captive specimen
25–30 cm

photo: G. COSENTINO

RT05791-4 *Chelodina (Chelodina) pritchardi* RHODIN, 1994
Moitaka, Central Prov., Papua New Guinea
18–25 cm

photo: M. O'SHEA

RT05792-4 *Chelodina (Chelodina) pritchardi* RHODIN, 1994
Moitaka, Central Prov., Papua New Guinea
18–25 cm

photo: M. O'SHEA

Chelodina (Chelodina) pritchardi (red),
Chelodina (Chelodina) reimanni (blue)

RT05801-4 *Chelodina (Chelodina) reimanni* PHILIPPEN & GROSSMANN, 1990
Merauke Regency, Papua, Indonesia
16–21 cm
♀
photo: C. HAGEN

35°C 3

RT05802-4 *Chelodina (Chelodina) reimanni* PHILIPPEN & GROSSMANN, 1990
Captive specimen
16–21 cm
♀
photo: S. SZYMANSKI

35°C 3

RT05803-4 *Chelodina (Chelodina) reimanni* PHILIPPEN & GROSSMANN, 1990
Merauke Regency, Papua, Indonesia
16–21 cm
♀
photo: C. HAGEN

35°C 3

RT05804-4 *Chelodina (Chelodina) reimanni* PHILIPPEN & GROSSMANN, 1990
Captive specimen
16–21 cm
♂
photo: S. SZYMANSKI

35°C 3

RT05805-4 *Chelodina (Chelodina) reimanni* PHILIPPEN & GROSSMANN, 1990
Merauke Regency, Papua, Indonesia
16–21 cm
photo: C. HAGEN

35°C 3

RT05806-1 *Chelodina (Chelodina) reimanni* PHILIPPEN & GROSSMANN, 1990
16–21 cm
Hatching
photo: S. SZYMANSKI

35°C 3

RT05811-4 *Chelodina (Chelodina) steindachneri* SIEBENROCK, 1914
Murchison River, WA, Australia
15–19 cm

photo: G. SCHMIDA

35°C 3

RT05812-4 *Chelodina (Chelodina) steindachneri* SIEBENROCK, 1914
Karrara Rangelands, WA, Australia
15–19 cm

photo: J. LOCHMAN/LOCHMAN LT

35°C 3

RT05813-4 *Chelodina (Chelodina) steindachneri* SIEBENROCK, 1914
Wondgina, WA, Australia
15–19 cm

photo: R. SOMAWEERA

35°C 3

RT05814-4 *Chelodina (Chelodina) steindachneri* SIEBENROCK, 1914
Wondgina, WA, Australia
15–19 cm

photo: R. SOMAWEERA

35°C 3

RT05815-3 *Chelodina (Chelodina) steindachneri* SIEBENROCK, 1914
Nullagine River, WA, Australia
15–19 cm
Subadult

photo: D. HEGNER

35°C 3

RT05816-3 *Chelodina (Chelodina) steindachneri* SIEBENROCK, 1914
Nullagine River, WA, Australia
15–19 cm
Subadult

photo: D. HEGNER

35°C 3

RT05817-4 *Chelodina (Chelodina) steindachneri* SIEBENROCK, 1914
Fortescue River, WA, Australia
15–19 cm

photo: A. ELLIOTT

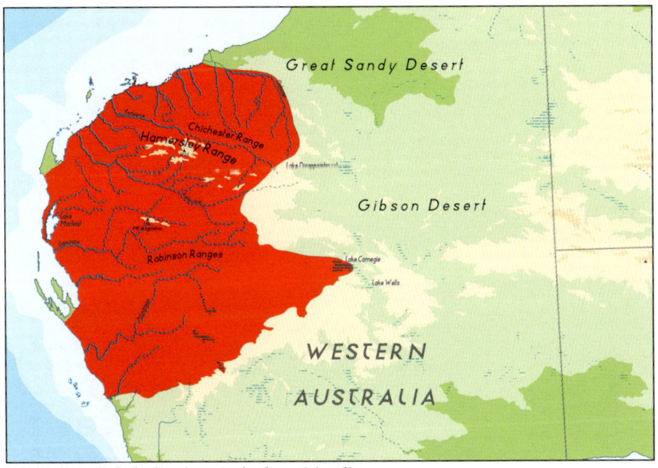

Chelodina (Chelodina) steindachneri (red)

RT05818-4 *Chelodina (Chelodina) steindachneri* SIEBENROCK, 1914
Fortescue River, WA, Australia
15–19 cm

photo: A. ELLIOTT

RT05821-4 *Chelodina (Macrochelodina) expansa* GRAY, 1857
Fitzroy River, QLD, Australia
40–48 cm

photo: J. CANN

RT05822-4 *Chelodina (Macrochelodina) expansa* GRAY, 1857
Fitzroy River, QLD, Australia
40–48 cm

photo: J. CANN

RT05823-4 *Chelodina (Macrochelodina) expansa* GRAY, 1857
Logan River, QLD, Australia
40–48 cm

photo: G. SCHMIDA

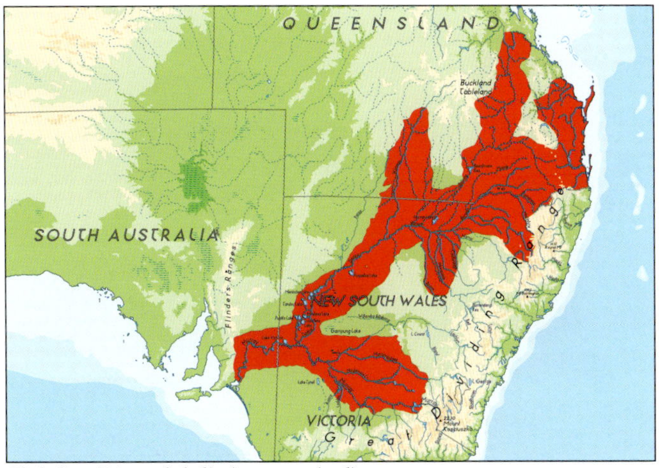

Chelodina (*Macrochelodina*) *expansa* (red)

RT05824-4 *Chelodina* (*Macrochelodina*) *expansa* GRAY, 1857
Logan River, QLD, Australia
40–48 cm

photo: G. SCHMIDA

35°C 3

RT05825-5 *Chelodina* (*Macrochelodina*) *expansa* GRAY, 1857
Fraser Island, QLD, Australia
40–48 cm

photo: J. CANN

35°C 3

RT05826-4 *Chelodina* (*Macrochelodina*) *expansa* GRAY, 1857
Brisbane River, QLD, Australia
40–48 cm

photo: C. LATTA

35°C 3

RT05827-4 *Chelodina* (*Macrochelodina*) *expansa* GRAY, 1857
Fraser Island, QLD, Australia
40–48 cm

photo: J. CANN

35°C 3

RT05828 *Chelodina* (*Macrochelodina*) *expansa* GRAY, 1857
Habitat, Lake Mckenzie, Fraser Island, QLD, Australia

photo: G. BROWN

RT05829-4 *Chelodina* (*Macrochelodina*) *expansa* GRAY, 1857
Lake Allom, Fraser Island, QLD, Australia
40–48 cm

photo: R. CLARKE & J. L. MCKAY

RT05830-4 *Chelodina* (*Macrochelodina*) *expansa* GRAY, 1857
Lake Allom, Fraser Island, QLD, Australia
40–48 cm

photo: R. CLARKE & L. MCKAY

RT05841-4 *Chelodina* (*Macrochelodina*) *kuchlingi* CANN, 1997
Holotype (WAM R29411)
20–24 cm

photo: G. KUCHLING

RT05842-4 *Chelodina* (*Macrochelodina*) *kuchlingi* CANN, 1997
Holotype (WAM R29411)
20–24 cm

photo: G. KUCHLING

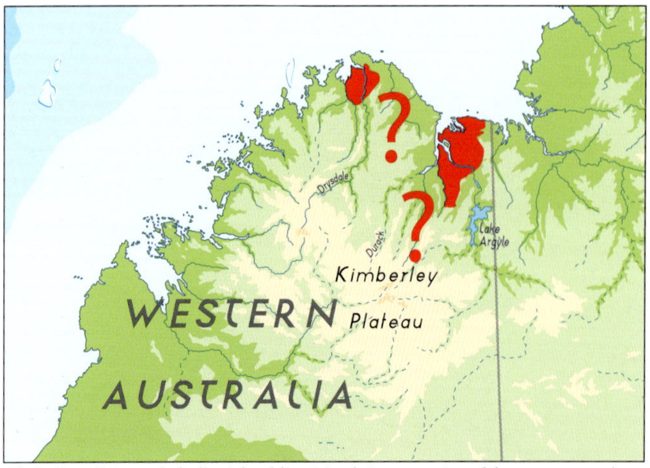

Chelodina (*Macrochelodina*) *kuchlingi* (red; ? – questionable occurrence)

RT05851-4 *Chelodina* (*Macrochelodina*) *oblonga oblonga* GRAY, 1841
Darwin, NT, Australia
22–28 cm

photo: R. SOMAWEERA

RT05852-4 *Chelodina* (*Macrochelodina*) *oblonga oblonga* Gray, 1841
Daly River, NT, Australia
22–28 cm

photo: J. Cann

RT05853 *Chelodina* (*Macrochelodina*) *oblonga oblonga* Gray, 1841
Habitat near Darwin, NT, Australia

photo: R. Somaweera

RT05854-4 *Chelodina* (*Macrochelodina*) *oblonga oblonga* Gray, 1841
Daly River, NT, Australia
22–28 cm

photo: J. Cann

RT05855-4 *Chelodina* (*Macrochelodina*) *oblonga oblonga* GRAY, 1841
South Alligator River, Kakadu NP, NT, Australia
22–28 cm

photo: J. L. MCKAY

RT05856-4 *Chelodina* (*Macrochelodina*) *oblonga oblonga* GRAY, 1841
South Alligator River, Kakadu NP, NT, Australia
22–28 cm

photo: J. L. MCKAY

Chelodina (*Macrochelodina*) *oblonga oblonga* (red),
Chelodina (*Macrochelodina*) *oblonga burrungandjii* (blue),
Chelodina (*Macrochelodina*) *oblonga rugosa* (green),
Chelodina (*Macrochelodina*) *oblonga walloyarrina* (orange),
area of intergradation (pink)

RT05857-4 *Chelodina* (*Macrochelodina*) *oblonga oblonga* GRAY, 1841
Darwin, NT, Australia
22–28 cm

photo: G. SCHMIDA

RT05858-4 *Chelodina* (*Macrochelodina*) *oblonga oblonga* GRAY, 1841
Darwin, NT, Australia
22–28 cm

photo: G. SCHMIDA

RT05859-4 *Chelodina* (*Macrochelodina*) *oblonga oblonga* Gray, 1841
Darwin, NT, Australia
22–28 cm

photo: G. Schmida

RT05861-4 *Chelodina* (*Macrochelodina*) *oblonga burrungandjii* Thomson, Kennett & Georges, 2000
Arnhem Land, NT, Australia
31–36 cm

photo: J. Cann

RT05862-4 *Chelodina* (*Macrochelodina*) *oblonga burrungandjii* Thomson, Kennett & Georges, 2000
Arnhem Land, NT, Australia
31–36 cm

photo: G. Schmida

RT05863-4 *Chelodina (Macrochelodina) oblonga burrungandjii* THOMSON, KENNETT & GEORGES, 2000
Arnhem Land, NT, Australia
31–36 cm
photo: G. SCHMIDA

RT05864-4 *Chelodina (Macrochelodina) oblonga burrungandjii* THOMSON, KENNETT & GEORGES, 2000
South Alligator River drainage, NT, Australia
31–36 cm
photo: J. CANN

RT05865-4 *Chelodina (Macrochelodina) oblonga burrungandjii* THOMSON, KENNETT & GEORGES, 2000
Arnhem Land, NT, Australia
31–36 cm
photo: G. SCHMIDA

RT05871-4 *Chelodina* (*Macrochelodina*) *oblonga rugosa* OGILBY, 1890
Merauke Regency, Papua, Indonesia
25–40 cm
♀
photo: C. HAGEN

RT05872-4 *Chelodina* (*Macrochelodina*) *oblonga rugosa* OGILBY, 1890
Merauke Regency, Papua, Indonesia
25–40 cm
♀
photo: C. HAGEN

RT05873-4 *Chelodina* (*Macrochelodina*) *oblonga rugosa* OGILBY, 1890
Merauke Regency, Papua, Indonesia
25–40 cm
♀
photo: C. HAGEN

RT05881-4 *Chelodina* (*Macrochelodina*) *oblonga walloyarrina* (McCord & Joseph-Ouni, 2007)
Little Mertens Falls, Mitchell Plateau, WA, Australia
24–30 cm
photo: J. L. McKay

RT05882-4 *Chelodina* (*Macrochelodina*) *oblonga walloyarrina* (McCord & Joseph-Ouni, 2007)
Little Mertens Falls, Mitchell Plateau, WA, Australia
24–30 cm
photo: J. L. McKay

RT05883-4 *Chelodina* (*Macrochelodina*) *oblonga walloyarrina* (McCord & Joseph-Ouni, 2007)
Amalia Gorge, El Questro, WA, Australia
24–30 cm
photo: S. Wilson

RT05884-4 *Chelodina* (*Macrochelodina*) *oblonga walloyarrina* (McCord & Joseph-Ouni, 2007)
Mitchell Plateau, Kimberley, WA, Australia
24–30 cm
photo: J. Lochman/Lochman LT

RT05885-4 *Chelodina* (*Macrochelodina*) *oblonga walloyarrina* (McCord & Joseph-Ouni, 2007)
Fitzroy River, WA, Australia
24–30 cm
photo: J. Cann

RT05886-2 *Chelodina* (*Macrochelodina*) *oblonga walloyarrina* (McCord & Joseph-Ouni, 2007)
Carson River, Kimberley, WA, Australia
24–30 cm, Juvenile
photo: G. Schmida

RT05887-4 *Chelodina* (*Macrochelodina*) *oblonga walloyarrina* (McCord & Joseph-Ouni, 2007)
Fitzroy River, WA, Australia
24–30 cm
photo: J. Cann

35°C 3

RT05888-2 *Chelodina* (*Macrochelodina*) *oblonga walloyarrina* (McCord & Joseph-Ouni, 2007)
Carson River, Kimberley, WA, Australia
24–30 cm, Juvenile
photo: G. Schmida

35°C 3

RT05891-4 *Chelodina (Macrochelodina) parkeri* RHODIN & MITTERMEIER, 1976
Papua New Guinea
22–30 cm

photo: M. O'SHEA

35°C 3

RT05892-4 *Chelodina (Macrochelodina) parkeri* RHODIN & MITTERMEIER, 1976
Papua New Guinea
22–30 cm

photo: M. O'SHEA

35°C 3

RT05893-4 *Chelodina (Macrochelodina) parkeri* RHODIN & MITTERMEIER, 1976
Captive specimen
22–30 cm
♂

photo: H. PROKOP

35°C 3

Chelodina (Macrochelodina) parkeri (red)

RT05894-4 *Chelodina (Macrochelodina) parkeri* RHODIN & MITTERMEIER, 1976
Captive specimen
22–30 cm
♀

photo: H. PROKOP

35°C 3

RT05895-4 *Chelodina (Macrochelodina) parkeri* Rhodin & Mittermeier, 1976
Captive specimen
22–30 cm
♂
photo: H. Prokop

RT05896-4 *Chelodina (Macrochelodina) parkeri* Rhodin & Mittermeier, 1976
Captive specimen
22–30 cm
♀
photo: H. Prokop

RT05897-4 *Chelodina (Macrochelodina) parkeri* Rhodin & Mittermeier, 1976
Papua New Guinea
22–30 cm
photo: M. O'Shea

RT05901-4 *Chelodina* (*Macrodiremys*) *colliei* GRAY, 1856
Perth, WA, Australia
25–31 cm

photo: G. SCHMIDA

RT05902-4 *Chelodina* (*Macrodiremys*) *colliei* GRAY, 1856
Thomson's Lake Nature Reserve, Perth, WA, Australia
25–31 cm

photo: D. BETTINI/LOCHMAN LT

RT05903-4 *Chelodina* (*Macrodiremys*) *colliei* GRAY, 1856
Brookton, WA, Australia
25–31 cm

photo: A. ELLIOTT

RT05904-4 *Chelodina* (*Macrodiremys*) *colliei* GRAY, 1856
Perth, WA, Australia
25–31 cm

photo: G. KUCHLING

RT05905-4 *Chelodina (Macrodiremys) colliei* Gray, 1856
Near Pemberton, WA, Australia
25–31 cm

photo: J. Lochman/Lochman LT

35°C 3

RT05906-4 *Chelodina (Macrodiremys) colliei* Gray, 1856
Brookton, WA, Australia
25–31 cm

photo: A. Elliott

35°C 3

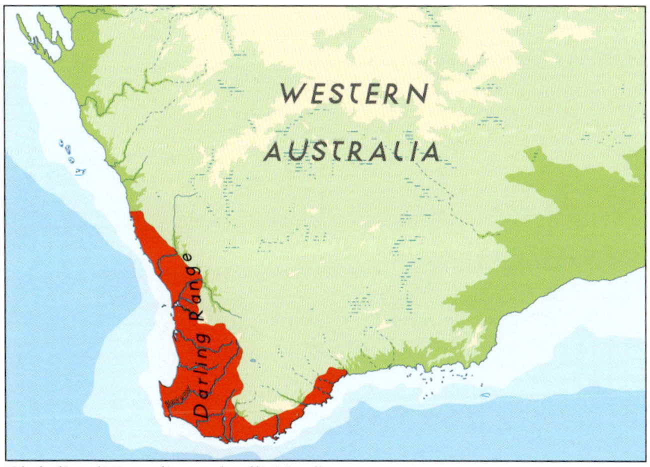

Chelodina (Macrodiremys) colliei (red)

RT05907-4 *Chelodina (Macrodiremys) colliei* Gray, 1856
Perth, WA, Australia
25–31 cm

photo: G. Schmida

RT05908-4 *Chelodina (Macrodiremys) colliei* Gray, 1856
Southwest WA, Australia
25–31 cm

photo: J. Cann

RT05909-4 *Chelodina (Macrodiremys) colliei* Gray, 1856
Perth, WA, Australia
25–31 cm

photo: G. Kuchling

RT05910-4 *Chelodina (Macrodiremys) colliei* GRAY, 1856
Blue Gum Lake Reserve, Perth, WA, Australia
25–31 cm

photo: J. LOCHMAN/LOCHMAN LT

RT05911-2 *Chelodina (Macrodiremys) colliei* GRAY, 1856
Captive specimen
25–31 cm
Juvenile

photo: G. SCHMIDA

RT05912-4 *Chelodina (Macrodiremys) colliei* GRAY, 1856
Captive specimen
25–31 cm

photo: R. MIDTGAARD

RT05921-4 *Elseya (Elseya) branderhorsti* (OUWENS, 1914)
Bensbach River, Western Prov., Papua New Guinea
30–43 cm
photo: M. O'SHEA

RT05922-4 *Elseya (Elseya) branderhorsti* (OUWENS, 1914)
Bensbach River, Western Prov., Papua New Guinea
30–43 cm
photo: M. O'SHEA

RT05923-4 *Elseya (Elseya) branderhorsti* (OUWENS, 1914)
Bensbach River, Western Prov., Papua New Guinea
30–43 cm
photo: M. O'SHEA

RT05924-3 *Elseya (Elseya) branderhorsti* (OUWENS, 1914)
Bian River, Papua, Indonesia
30–43 cm
Subadult
photo: C. HAGEN

RT05925-3 *Elseya (Elseya) branderhorsti* (OUWENS, 1914)
Bian River, Papua, Indonesia
30–43 cm
Subadult
photo: C. HAGEN

RT05926-2 *Elseya (Elseya) branderhorsti* (OUWENS, 1914)
Captive specimen
30–43 cm
Juvenile
photo: H. PROKOP

RT05927 *Elseya (Elseya) branderhorsti* (Ouwens, 1914)
Habitat, Bian River, Papua, Indonesia

photo: C. Hagen

Elseya (Elseya) branderhorsti (red),
Elseya (Elseya) dentata (blue),
Elseya (Elseya) flaviventralis (green)

RT05931-4 *Elseya (Elseya) dentata* (Gray, 1863)
Carson River, WA, Australia
28–33 cm

photo: J. Cann

35°C 3

RT05932-2 *Elseya (Elseya) dentata* (Gray, 1863)
Carson River, WA, Australia
28–33 cm
Juvenile

photo: G. Schmida

RT05933 *Elseya (Elseya) dentata* (Gray, 1863)
Habitat, Lake Argyle, WA, Australia

photo: R. Somaweera

RT05934-4 *Elseya (Elseya) dentata* (Gray, 1863)
Daly River, NT, Australia
28–33 cm
♂

photo: G. Schmida/Lochman LT

RT05935-4 *Elseya (Elseya) dentata* (Gray, 1863)
Roper River, NT, Australia
28–33 cm
♀
photo: J. Cann

RT05936-4 *Elseya (Elseya) dentata* (Gray, 1863)
Daly River, NT, Australia
28–33 cm
♀
photo: G. Schmida

RT05937-3 *Elseya (Elseya) dentata* (Gray, 1863)
Roper River, NT, Australia
28–33 cm
Subadult
photo: J. Cann

RT05938-4 *Elseya (Elseya) dentata* (Gray, 1863)
Daly drainage, NT, Australia
28–33 cm
photo: J. Cann

RT05940-2 *Elseya (Elseya) dentata* (Gray, 1863)
Roper River, NT, Australia
28–33 cm
Juvenile
photo: J. Cann

RT05941-1 *Elseya (Elseya) dentata* (Gray, 1863)
Captive specimen
28–33 cm
Hatchling
photo: A. Elliott

RT05942-5 *Elseya (Elseya) dentata* (Gray, 1863)
Koolpin Gorge, Kakadu NP, NT, Australia
28–33 cm

photo: R. Clarke & J. L. McKay

RT05951-2 *Elseya (Elseya) flaviventralis* Thomson & Georges, 2016
Alligator River, NT, Australia
27–34 cm
Juvenile

photo: J. Cann

RT05952-4 *Elseya (Elseya) flaviventralis* Thomson & Georges, 2016
Alligator River, NT, Australia
27–34 cm

photo: J. Cann

RT05953-4 *Elseya (Elseya) flaviventralis* Thomson & Georges, 2016
Alligator River, NT, Australia
27–34 cm

photo: J. Cann

RT05961-4 *Elseya (Hanwarachelys) novaeguineae* (MEYER, 1874)
Kaimana Bird's Head Peninsula, West Papua, Indonesia
20–26 cm

photo: J. CANN courtesy of W.P. MCCORD

RT05962-4 *Elseya (Hanwarachelys) novaeguineae* (MEYER, 1874)
Kaimana Bird's Head Peninsula, West Papua, Indonesia
20–26 cm

photo: J. CANN courtesy of W.P. MCCORD

RT05963-4 *Elseya (Hanwarachelys) novaeguineae* (MEYER, 1874)
Kaimana Bird's Head Peninsula, West Papua, Indonesia
20–26 cm

photo: J. CANN courtesy of W.P. MCCORD

Elseya (Hanwarachelys) novaeguineae (red; * – introduced populations),
Elseya (Hanwarachelys) rhodini (blue),
Elseya (Hanwarachelys) schultzei (green; ? – questionable occurrence;
* – introduced populations)

RT05971-3 *Elseya (Hanwarachelys) rhodini* THOMSON, AMEPOU, ANAMIATO & GEORGES, 2015
Merauke Regency, Papua, Indonesia
20–26 cm, Subadult
photo: C. HAGEN

RT05972-4 *Elseya (Hanwarachelys) rhodini* THOMSON, AMEPOU, ANAMIATO & GEORGES, 2015
Girigarande, Western Prov., Papua New Guinea
20–26 cm
photo: M. O'SHEA

RT05973-4 *Elseya (Hanwarachelys) rhodini* THOMSON, AMEPOU, ANAMIATO & GEORGES, 2015
Girigarande, Western Prov., Papua New Guinea
20–26 cm
photo: M. O'SHEA

RT05974-4 *Elseya (Hanwarachelys) rhodini* THOMSON, AMEPOU, ANAMIATO & GEORGES, 2015
Girigarande, Western Prov., Papua New Guinea
20–26 cm
photo: M. O'SHEA

RT05981-4 *Elseya (Hanwarachelys) schultzei* (VOGT, 1911)
Madang Prov., Papua New Guinea
20–26 cm
photo: M. O'SHEA

RT05982-4 *Elseya (Hanwarachelys) schultzei* (Vogt, 1911)
Northern Papua New Guinea
20–26 cm
♂
photo: H. Prokop

RT05983-4 *Elseya (Hanwarachelys) schultzei* (Vogt, 1911)
Northern Papua New Guinea
20–26 cm
♂ + ♀
photo: H. Prokop

RT05984-4 *Elseya (Hanwarachelys) schultzei* (Vogt, 1911)
Northern Papua New Guinea
20–26 cm
♂
photo: H. Prokop

RT05991-4 *Elseya (Pelocomastes) albagula* THOMSON, GEORGES & LIMPUS, 2006
Mary River, QLD, Australia
28–42 cm

photo: J. CANN

35°C 3

RT05992-4 *Elseya (Pelocomastes) albagula* THOMSON, GEORGES & LIMPUS, 2006
Mary River, QLD, Australia
28–42 cm

photo: J. CANN

35°C 3

RT05993-4 *Elseya (Pelocomastes) albagula* THOMSON, GEORGES & LIMPUS, 2006
Burnett River, QLD, Australia
28–42 cm

photo: G. SCHMIDA

35°C 3

RT05994-4 *Elseya (Pelocomastes) albagula* THOMSON, GEORGES & LIMPUS, 2006
Burnett River, QLD, Australia
28–42 cm

photo: G. SCHMIDA

RT05995-4 *Elseya (Pelocomastes) albagula* THOMSON, GEORGES & LIMPUS, 2006
Mary River, QLD, Australia
28–42 cm

photo: C. LATTA

RT05996-4 *Elseya (Pelocomastes) albagula* THOMSON, GEORGES & LIMPUS, 2006
Captive specimen
28–42 cm

photo: S. WILSON

RT05997-4 *Elseya (Pelocomastes) albagula* THOMSON, GEORGES & LIMPUS, 2006
Connors River, QLD, Australia
28–42 cm

photo: A. ELLIOTT

RT05998-2 *Elseya (Pelocomastes) albagula* THOMSON, GEORGES & LIMPUS, 2006
Mary River, QLD, Australia
28–42 cm
Juvenile

photo: J. CANN

RT05999-2 *Elseya (Pelocomastes) albagula* THOMSON, GEORGES & LIMPUS, 2006
Mary River, QLD, Australia
28–42 cm
Juvenile

photo: G. SCHMIDA

RT06000-4 *Elseya (Pelocomastes) albagula* THOMSON, GEORGES & LIMPUS, 2006
Connors River, QLD, Australia
28–42 cm

photo: A. ELLIOTT

35°C 3

RT06001 *Elseya (Pelocomastes) albagula* THOMSON, GEORGES & LIMPUS, 2006
Habitat, Connors River, QLD, Australia

photo: A. ELLIOTT

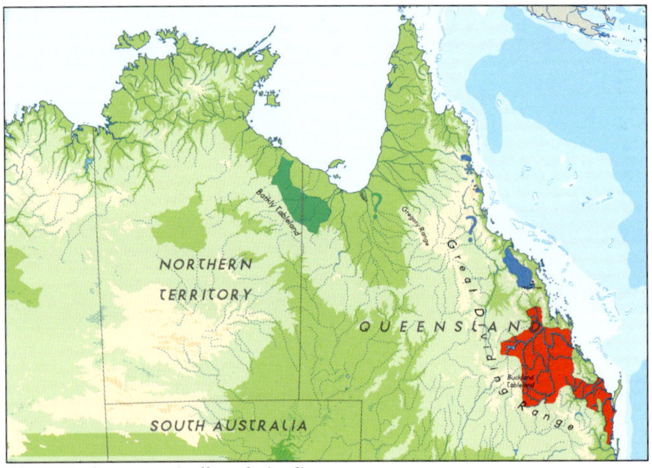

Elseya (Pelocomastes) albagula (red),
Elseya (Pelocomastes) irwini (blue; ? – questionable occurrence),
Elseya (Pelocomastes) lavarackorum (green; ? – questionable occurrence)

RT06011-4 *Elseya (Pelocomastes) irwini* CANN, 1997
Broken River, QLD, Australia
25–35 cm

photo: G. SCHMIDA

35°C 3

RT06012-4 *Elseya (Pelocomastes) irwini* CANN, 1997

25–35 cm

photo: C. LATTA

35°C 3

RT06013-4 *Elseya (Pelocomastes) irwini* CANN, 1997
Burdekin River drainage, QLD, Australia
25–35 cm

photo: J. CANN

35°C 3

RT06014-4 *Elseya (Pelocomastes) irwini* CANN, 1997
Broken River, QLD, Australia
25–35 cm

photo: G. SCHMIDA

RT06015-4 *Elseya (Pelocomastes) irwini* CANN, 1997
Broken River, QLD, Australia
25–35 cm

photo: C. LATTA

RT06016-4 *Elseya (Pelocomastes) irwini* CANN, 1997

25–35 cm

photo: C. LATTA

RT06021-4 *Elseya* (*Pelocomastes*) *lavarackorum* (White & Archer, 1994)
Gregory River, QLD, Australia
25–32 cm
♂
photo: G. Schmida

RT06022-4 *Elseya* (*Pelocomastes*) *lavarackorum* (White & Archer, 1994)
Gregory River, QLD, Australia
25–32 cm
photo: G. Schmida

RT06023-4 *Elseya* (*Pelocomastes*) *lavarackorum* (White & Archer, 1994)
Gregory River, QLD, Australia
25–32 cm
♀
photo: G. Schmida

RT06024-4 *Elseya (Pelocomastes) lavarackorum* (White & Archer, 1994)
Gregory River, QLD, Australia
25–32 cm
♂
photo: G. Schmida

RT06025-4 *Elseya (Pelocomastes) lavarackorum* (White & Archer, 1994)
The Cascades, Boodjamulla NP, QLD, Australia
25–32 cm
photo: S. Wilson

RT06026-4 *Elseya (Pelocomastes) lavarackorum* (White & Archer, 1994)
Gregory River, QLD, Australia
25–32 cm
♂ + ♀
photo: G. Schmida

RT06031-5 *Elusor macrurus* Cann & Legler, 1994
Mary River, QLD, Australia
30–44 cm

photo: J. Cann

RT06032-4 *Elusor macrurus* Cann & Legler, 1994
Mary River, QLD, Australia
30–44 cm

photo: J. Cann

RT06033-4 *Elusor macrurus* Cann & Legler, 1994
Mary River, QLD, Australia
30–44 cm
♀

photo: G. Schmida

RT06034-4 *Elusor macrurus* Cann & Legler, 1994
Mary River, QLD, Australia
30–44 cm
♀
photo: G. Schmida

RT06035-4 *Elusor macrurus* Cann & Legler, 1994
Mary River, QLD, Australia
30–44 cm
photo: A. Elliott

RT06036-4 *Elusor macrurus* Cann & Legler, 1994
Mary River, QLD, Australia
30–44 cm
♂
photo: G. Schmida

Elusor macrurus (red)

RT06037-1 *Elusor macrurus* Cann & Legler, 1994
Mary River, QLD, Australia
30–44 cm
Hatchling
photo: J. Cann

RT06041-4 *Emydura australis* (GRAY, 1841)
Lake Argyle, WA, Australia
16–23 cm

photo: R. SOMAWEERA

RT06042-4 *Emydura australis* (GRAY, 1841)
Lake Argyle, WA, Australia
16–23 cm

photo: R. SOMAWEERA

RT06043-4 *Emydura australis* (GRAY, 1841)
Bell Gorge, WA, Australia
16–23 cm

photo: J. LOCHMAN/LOCHMAN LT

RT06044-5 *Emydura australis* (Gray, 1841)
Bell Gorge, WA, Australia
16–23 cm

photo: J. Lochman/Lochman LT

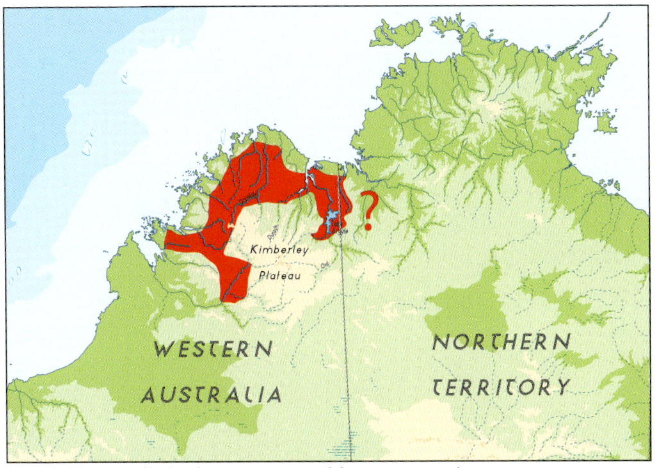

Emydura australis (red; ? – questionable occurrence)

RT06045-3 *Emydura australis* (Gray, 1841)
El Questro Station, WA, Australia
16–23 cm
Subadult

photo: S. Wilson

RT06046-4 *Emydura australis* (Gray, 1841)
Mitchell Plateau, WA, Australia
16–23 cm

photo: J. Lochman/Lochman LT

RT06047-5 *Emydura australis* (Gray, 1841)
Charnley River, WA, Australia
16–23 cm

photo: D. Bettini/Lochman LT

RT06048-2 *Emydura australis* (Gray, 1841)
Mitchell Plateau, WA, Australia
16–23 cm
Juvenile

photo: J. Lochman/Lochman LT

RT06051-4 *Emydura krefftii krefftii* (Gray, 1871)
Raglan Creek, QLD, Australia
24–35 cm

photo: G. Schmida

RT06052-4 *Emydura krefftii krefftii* (Gray, 1871)
Little Yabba Creek, QLD, Australia
24–35 cm

photo: A. Elliott

RT06053-4 *Emydura krefftii krefftii* (Gray, 1871)
Little Yabba Creek, QLD, Australia
24–35 cm

photo: A. Elliott

RT06054-4 *Emydura krefftii krefftii* (Gray, 1871)
Captive specimens
24–35 cm
♂ + ♀
photo: H. Prokop

RT06055-4 *Emydura krefftii krefftii* (Gray, 1871)
Burnett River, QLD, Australia
24–35 cm
photo: S. Wilson

RT06056-4 *Emydura krefftii krefftii* (Gray, 1871)
Raglan Creek, QLD, Australia
24–35 cm
photo: G. Schmida

RT06057-4 *Emydura krefftii krefftii* (Gray, 1871)
Mary River, QLD, Australia
24–35 cm
Subadult
photo: G. Schmida

RT06058-1 *Emydura krefftii krefftii* (Gray, 1871)
24–35 cm
Hatching
photo: D. Goodall

RT06059-2 *Emydura krefftii krefftii* (Gray, 1871)
Mary River, QLD, Australia
24–35 cm
Juvenile
photo: G. Schmida

RT06060-2 *Emydura krefftii krefftii* (GRAY, 1871)
Mary River, QLD, Australia
24–35 cm
Juvenile
photo: G. SCHMIDA

RT06061-2 *Emydura krefftii krefftii* (GRAY, 1871)
Mary River, QLD, Australia
24–35 cm
Juvenile
photo: G. SCHMIDA

RT06071-4 *Emydura krefftii emmotti* CANN, MCCORD & JOSEPH-OUNI *in* MCCORD, CANN & JOSEPH-OUNI , 2003
Thomson River, QLD, Australia
25–37 cm
photo: G. SCHMIDA

RT06072-5 *Emydura krefftii emmotti* CANN, MCCORD & JOSEPH-OUNI *in* MCCORD, CANN & JOSEPH-OUNI , 2003
Lake Dunn, Thomson River drainage, QLD, Australia
25–37 cm
photo: J. CANN

RT06073-5 *Emydura krefftii emmotti* CANN, MCCORD & JOSEPH-OUNI *in* MCCORD, CANN & JOSEPH-OUNI , 2003
Thomson River, QLD, Australia
25–37 cm
photo: J. CANN

RT06074-4 *Emydura krefftii emmotti* CANN, MCCORD & JOSEPH-OUNI *in* MCCORD, CANN & JOSEPH-OUNI , 2003
Thomson River, QLD, Australia
25–37 cm
photo: G. SCHMIDA

RT06075-4 *Emydura krefftii emmotti* Cann, McCord & Joseph-Ouni *in* McCord, Cann & Joseph-Ouni, 2003
Thomson River, QLD, Australia
25–37 cm

photo: A. Elliott

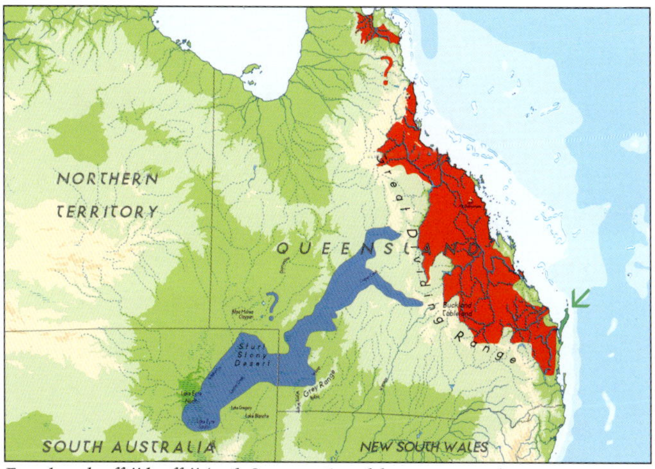

Emydura krefftii krefftii (red; ? – questionable occurrence), *Emydura krefftii emmotti* (blue; ? – questionable occurrence), *Emydura krefftii nigra* (green; see arrow)

RT06076-4 *Emydura krefftii emmotti* Cann, McCord & Joseph-Ouni *in* McCord, Cann & Joseph-Ouni, 2003
Thomson River, QLD, Australia
25–37 cm

photo: G. Schmida

RT06081-4 *Emydura krefftii nigra* McCord, Cann & Joseph-Ouni, 2003
Boomerang Lake, Fraser Island, QLD, Australia
20–25 cm

photo: J. Cann

35°C 3

RT06082-4 *Emydura krefftii nigra* McCord, Cann & Joseph-Ouni, 2003
Lake McKenzie, Fraser Island, QLD, Australia
20–25 cm

photo: R. Clarke & J. L. McKay

35°C 3

RT06083-4 *Emydura krefftii nigra* McCord, Cann & Joseph-Ouni, 2003
Fraser Island, QLD, Australia
20–25 cm

photo: J. Cann

35°C 3

RT06084-1 *Emydura krefftii nigra* McCord, Cann & Joseph-Ouni, 2003
Fraser Island, QLD, Australia
20–25 cm
Hatchling

photo: J. Cann

35°C 3

RT06091-4 *Emydura macquarii macquarii* (Gray, 1830)
Storm King Dam outflow, QLD, Australia
26–34 cm

photo: S. Wilson

35°C 3

RT06092-4 *Emydura macquarii macquarii* (Gray, 1830)
Lake Numalla, Paroo River catchment, QLD, Australia
26–34 cm

photo: G. Schmida

35°C 3

RT06093-4 *Emydura macquarii macquarii* (Gray, 1830)
Lake Numalla, Paroo River catchment, QLD, Australia
26–34 cm

photo: G. Schmida

35°C 3

RT06094-4 *Emydura macquarii macquarii* (Gray, 1830)
Lake Numalla, Paroo River catchment, QLD, Australia
26–34 cm

photo: G. Schmida

35°C 3

RT06095-2 *Emydura macquarii macquarii* (Gray, 1830)
Storm King Dam, QLD, Australia
20–21 cm
Juvenile

photo: J. Cann

35°C 3

RT06096 *Emydura macquarii macquarii* (GRAY, 1830)
Habitat, Darling River, near Wilcannia, NSW, Australia

photo: A. ELLIOTT

RT06101-4 *Emydura macquarii binjing* CANN, 1998
Grafton, NSW, Australia
20–21 cm

photo: A. ELLIOTT

RT06102-4 *Emydura macquarii binjing* CANN, 1998
Clarence River drainage, NSW, Australia
20–21 cm

photo: J. CANN

RT06103-4 *Emydura macquarii binjing* CANN, 1998
Grafton, NSW, Australia
20–21 cm

photo: A. ELLIOTT

RT06104-4 *Emydura macquarii binjing* CANN, 1998
Clarence River drainage, NSW, Australia
20–21 cm

photo: J. CANN

RT06111-4 *Emydura macquarii dharra* CANN, 1998
Macleay River, NSW, Australia
15–19 cm

photo: J. CANN

RT06112-4 *Emydura macquarii dharra* CANN, 1998
Captive specimens
15–19 cm
♂ + ♀

photo: H. PROKOP

RT06113-4 *Emydura macquarii dharra* CANN, 1998
Captive specimen
15–19 cm
♀

photo: H. PROKOP

RT06114-2 *Emydura macquarii dharra* CANN, 1998

15–19 cm
Juvenile

photo: A. ELLIOTT

RT06115-1 *Emydura macquarii dharra* CANN, 1998

15–19 cm
Hatching

photo: A. ELLIOTT

RT06116-4 *Emydura macquarii dharra* CANN, 1998
Macleay River, NSW, Australia
15–19 cm

photo: A. ELLIOTT

RT06121-4 *Emydura macquarii dharuk* CANN, 1998
Nepean River drainage, NSW, Australia
22–25 cm
♂

photo: J. CANN

35°C 3

RT06122-4 *Emydura macquarii dharuk* CANN, 1998
Nepean River drainage, NSW, Australia
22–25 cm
♀

photo: J. CANN

35°C 3

RT06123-4 *Emydura macquarii dharuk* CANN, 1998
Gosford, NSW, Australia
22–25 cm

photo: M. O'SHEA

35°C 3

RT06131-2 *Emydura macquarii gunabarra* CANN, 1998

23–26 cm
Juvenile

photo: A. ELLIOTT

35°C 3

RT06132-4 *Emydura macquarii gunabarra* CANN, 1998

23–26 cm

photo: A. ELLIOTT

35°C 3

RT06141-4 *Emydura macquarii signata* AHL, 1932
Lake Manchester, QLD, Australia
25–28 cm

photo: A. ELLIOTT

RT06142-4 *Emydura macquarii signata* AHL, 1932
Canungra, QLD, Australia
25–28 cm

photo: J. L. MCKAY

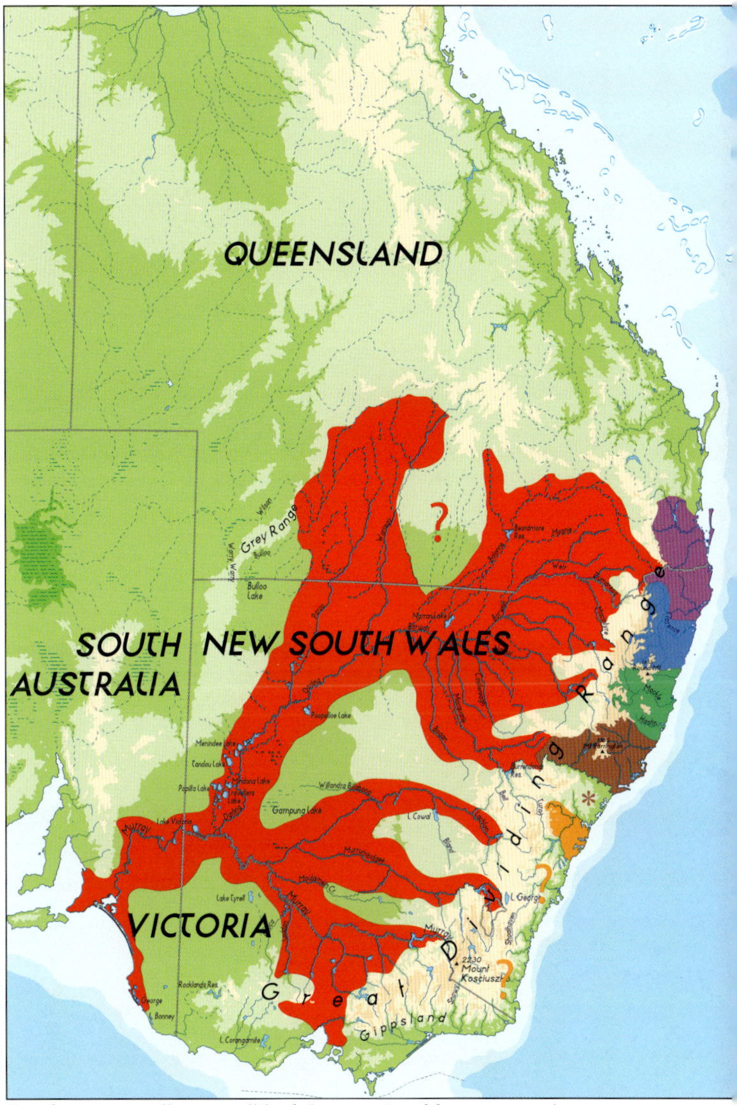

Emydura macquarii macquarii (red; ? – questionable occurrence),
Emydura macquarii binjing (blue),
Emydura macquarii dharra (green),
Emydura macquarii dharuk (orange; ? – questionable occurrence),
Emydura macquarii gunabarra (brown; * – introduced populations),
Emydura macquarii signata (purple)

RT06143-4 *Emydura macquarii signata* AHL, 1932
Brisbane River drainage, QLD, Australia
25–28 cm

photo: J. CANN

RT06144-4 *Emydura macquarii signata* AHL, 1932
Brisbane River, QLD, Australia
25–28 cm

photo: G. SCHMIDA

RT06145-4 *Emydura macquarii signata* AHL, 1932
Wynnum, QLD, Australia
25–28 cm

photo: S. WILSON

m ≋ ≅ ≡ 35°C 3

RT06146-4 *Emydura macquarii signata* AHL, 1932
QLD, Australia
25–28 cm

photo: A. ELLIOTT

m ≋ ≅ ≡ 35°C 3

RT06147-4 *Emydura macquarii signata* AHL, 1932
Brisbane River at Karana Downs, QLD, Australia
25–28 cm

photo: H. & J. BESTE/LOCHMAN LT

m ≋ ≅ ≡ 35°C 3

RT06151-4 *Emydura subglobosa* (KREFFT, 1876)
Captive specimen
19–26 cm

photo: A. ELLIOTT

RT06152-4 *Emydura subglobosa* (KREFFT, 1876)
Captive specimen
19–26 cm

photo: J. HARDING

RT06153-4 *Emydura subglobosa* (KREFFT, 1876)
Captive specimen
19–26 cm

photo: A. ELLIOTT

RT06154-4 *Emydura subglobosa* (KREFFT, 1876)
Captive specimen
19–26 cm

photo: H. PROKOP

RT06155-4 *Emydura subglobosa* (KREFFT, 1876)
Wasur NP, Merauke Regency. Papua, Indonesia
19–26 cm

photo: C. HAGEN

RT06156-4 *Emydura subglobosa* (KREFFT, 1876)
Captive specimen
19–26 cm
♂
photo: G. SCHMIDA

RT06157-4 *Emydura subglobosa* (KREFFT, 1876)
Captive specimen
19–26 cm
♂
photo: G. SCHMIDA

RT06158-4 *Emydura subglobosa* (Krefft, 1876)
Captive specimen
19–26 cm
♂
photo: G. Schmida

RT06159-4 *Emydura subglobosa* (Krefft, 1876)
Moitaka, Central Prov., Papua New Guinea
19–26 cm
photo: M. O'Shea

RT06160-4 *Emydura subglobosa* (Krefft, 1876)
Papua New Guinea
19–26 cm
photo: M. O'Shea

RT06161-4 *Emydura subglobosa* (Krefft, 1876)
Oriomo River, Western Prov., Papua New Guinea
19–26 cm
photo: M. O'Shea

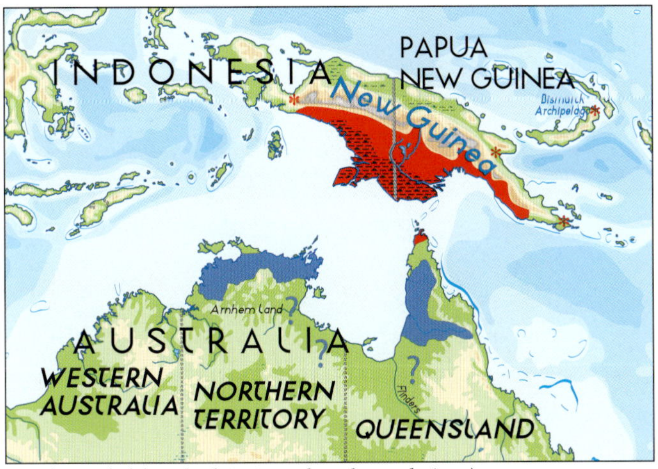

Emydura subglobosa (red; * – introduced populations), *Emydura tanybaraga* (blue; ? – questionable occurrence)

RT06162-1 *Emydura subglobosa* (Krefft, 1876)
National Capital District, Papua New Guinea
19–26 cm
Hatchling
photo: M. O'Shea

RT06171-4 *Emydura tanybaraga* CANN, 1997
South Alligator River, NT, Australia
21–29 cm

photo: G. SCHMIDA

35°C 3

RT06172-4 *Emydura tanybaraga* CANN, 1997
Finnis River, Darwin region, NT, Australia
21–29 cm

photo: G. SCHMIDA

35°C 3

RT06173-4 *Emydura tanybaraga* CANN, 1997
Finnis River, Darwin region, NT, Australia
21–29 cm

photo: G. SCHMIDA

35°C 3

RT06174-4 *Emydura tanybaraga* Cann, 1997
Darwin, NT, Australia
21–29 cm

photo: R. Somaweera

RT06175-2 *Emydura tanybaraga* Cann, 1997
Mataranka Hot Springs, NT, Australia
21–29 cm
Juvenile

photo: J. L. McKay

RT06176-4 *Emydura tanybaraga* Cann, 1997
Finnis River, Darwin region, NT, Australia
21–29 cm

photo: G. Schmida

RT06177-3 *Emydura tanybaraga* CANN, 1997
Lake Mitchell, QLD, Australia
21–29 cm
Subadult

photo: A. ELLIOTT

RT06178 *Emydura tanybaraga* CANN, 1997
Habitat, Mitchell River, QLD, Australia

photo: A. ELLIOTT

RT06179-4 *Emydura tanybaraga* CANN, 1997
Mitchell River, QLD, Australia
21–29 cm

photo: G. SCHMIDA

RT06181-4 *Emydura victoriae* (Gray, 1842)
Daly drainage, NT, Australia
19–25 cm

photo: J. Cann

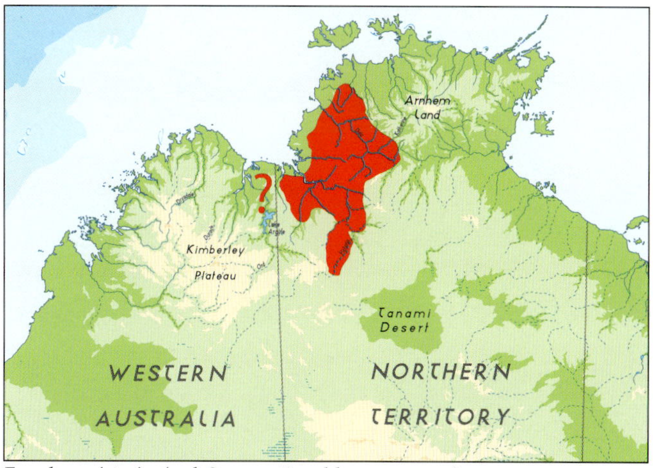

Emydura victoriae (red; ? – questionable occurrence)

RT06182-4 *Emydura victoriae* (Gray, 1842)
Gregory NP, NT, Australia
19–25 cm

photo: J. Lochman/Lochman LT

RT06183-4 *Emydura victoriae* (Gray, 1842)
Bullo River, Victoria River catchment, NT, Australia
19–25 cm

photo: G. Schmida

RT06184-4 *Emydura victoriae* (Gray, 1842)
Bullo River, Victoria River catchment, NT, Australia
19–25 cm

photo: G. Schmida

RT06185-4 *Emydura victoriae* (Gray, 1842)
Daly River, NT, Australia
19–25 cm

photo: S. Wilson

RT06186-4 *Emydura victoriae* (Gray, 1842)
Victoria River, NT, Australia
19–25 cm

photo: G. Schmida

RT06191-4 *Emydura worrelli* (Wells & Wellington, 1985)
Lawn Hill Creek, Nicholson River catchment, QLD, Australia
20–27 cm

photo: G. Schmida

RT06192-4 *Emydura worrelli* (Wells & Wellington, 1985)
Lawn Hill Creek, Nicholson River catchment, QLD, Australia
20–27 cm

photo: G. Schmida

RT06193-4 *Emydura worrelli* (Wells & Wellington, 1985)
Gregory River, Nicholson River catchment, QLD, Australia
20–27 cm

photo: G. Schmida

RT06194-3 *Emydura worrelli* (Wells & Wellington, 1985)
Gregory River, Nicholson River catchment, QLD, Australia
20–27 cm
Subadult

photo: J. Cann

RT06195-4 *Emydura worrelli* (Wells & Wellington, 1985)
Lawn Hill Creek, Nicholson River catchment, QLD, Australia
20–27 cm

photo: G. Schmida

RT06196-4 *Emydura worrelli* (Wells & Wellington, 1985)
Lawn Hill Creek, QLD, Australia
20–27 cm

photo: G. Schmida

RT06197-4 *Emydura worrelli* (Wells & Wellington, 1985)
Gregory River, Nicholson River catchment, QLD, Australia
20–27 cm

photo: G. Schmida

RT06198-4 *Emydura worrelli* (Wells & Wellington, 1985)
Lawn Hill Creek, QLD, Australia
20–27 cm

photo: C. Latta

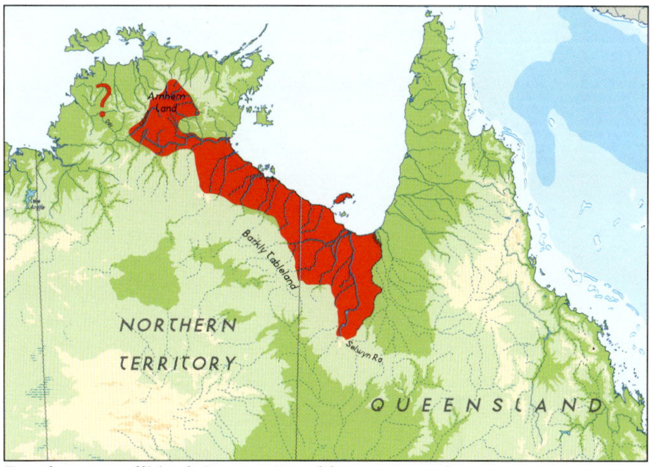

Emydura worrelli (red; ? – questionable occurrence)

RT06199-4 *Emydura worrelli* (Wells & Wellington, 1985)
Gregory River, QLD, Australia
20–27 cm

photo: G. Schmida

RT06200 *Emydura worrelli* (Wells & Wellington, 1985)
Habitat, Lawn Hill Gorge, QLD, Australia

photo: A. Elliott

RT06211-4 *Myuchelys bellii* (Gray, 1844)
Girraween NP, QLD, Australia
22–29 cm

photo: A. Elliott

35°C 3

RT06212-4 *Myuchelys bellii* (Gray, 1844)
Girraween NP, QLD, Australia
22–29 cm

photo: S. Wilson

35°C 3

RT06213-4 *Myuchelys bellii* (Gray, 1844)
Bald Rock Creek, QLD, Australia
22–29 cm

photo: G. Schmida

35°C 3

RT06214-4 *Myuchelys bellii* (GRAY, 1844)
Gwydir River, NSW, Australia
22–29 cm

photo: A. ELLIOTT

RT06215-4 *Myuchelys bellii* (GRAY, 1844)
Gwydir River, NSW, Australia
22–29 cm

photo: A. ELLIOTT

RT06216 *Myuchelys bellii* (GRAY, 1844)
Habitat, Gwydir River, NSW, Australia

photo: A. ELLIOTT

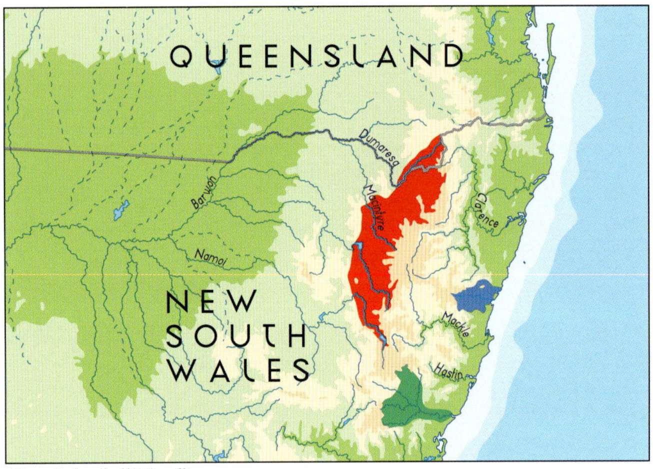

Myuchelys bellii (red),
Myuchelys georgesi (blue),
Myuchelys purvisi (green)

RT06221-4 *Myuchelys georgesi* (Cann, 1997)
Bellinger River, NSW, Australia
18–21 cm

photo: A. Elliott

RT06222-4 *Myuchelys georgesi* (Cann, 1997)
Bellinger River, NSW, Australia
18–21 cm

photo: A. Elliott

RT06223 *Myuchelys georgesi* (CANN, 1997)
Bellinger River, near Tora, NSW, Australia

photo: A. ELLIOTT

RT06224-4 *Myuchelys georgesi* (CANN, 1997)
Bellinger River, NSW, Australia
18–21 cm

photo: A. ELLIOTT

RT06231-5 *Myuchelys latisternum* (GRAY, 1867)
Jourama Falls, QLD, Australia
18–28 cm

photo: A. ELLIOTT

RT06232-4 *Myuchelys latisternum* (Gray, 1867)
Captive specimen
18–28 cm
♀
photo: H. Prokop

RT06233-4 *Myuchelys latisternum* (Gray, 1867)
Captive specimens
18–28 cm
♂+♀
photo: H. Prokop

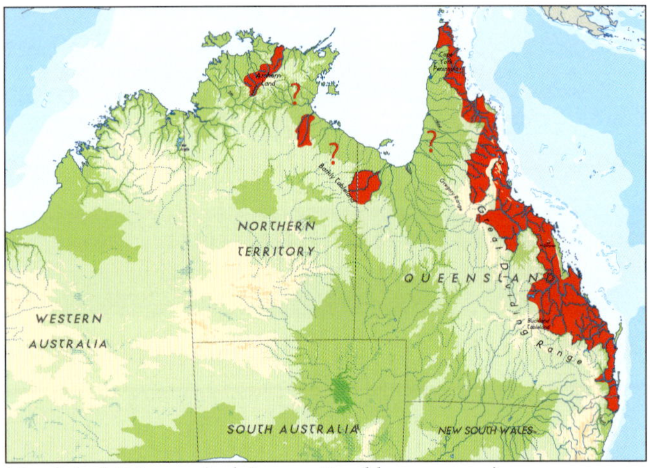

Myuchelys latisternum (red; ? – questionable occurrence)

RT06234-4 *Myuchelys latisternum* (Gray, 1867)
Jardine River, Cape York, QLD, Australia
18–28 cm
photo: J. Cann

RT06235-4 *Myuchelys latisternum* (Gray, 1867)
Lake Kurawongbah, QLD, Australia
18–28 cm
photo: S. Wilson

RT06236-4 *Myuchelys latisternum* (Gray, 1867)
Near Mount Elliot, QLD, Australia
18–28 cm
photo: A. Elliott

RT06237-4 *Myuchelys latisternum* (Gray, 1867)
Weipa region near the top of Cape York, QLD Australia
18–28 cm

photo: G. Schmida

35°C 3

RT06238-4 *Myuchelys latisternum* (Gray, 1867)
Gold Coast, QLD, Australia
18–28 cm

photo: G. Schmida

35°C 3

RT06239-4 *Myuchelys latisternum* (GRAY, 1867)
Gold Coast, QLD, Australia
18–28 cm
photo: G. SCHMIDA

RT06240-2 *Myuchelys latisternum* (GRAY, 1867)
Sheldon's Lagoon, Cape York, QLD, Australia
18–28 cm
Juvenile
photo: G. SCHMIDA

RT06241-2 *Myuchelys latisternum* (GRAY, 1867)
Maguk, Kakadu NP, NT, Australia
18–28 cm
Juvenile
photo: R. CLARKE & J. L. MCKAY

RT06242 *Myuchelys latisternum* (GRAY, 1867)
Habitat, QLD, Australia
photo: A. ELLIOTT

RT06243-2 *Myuchelys latisternum* (GRAY, 1867)
Maguk, Kakadu NP, NT, Australia
18–28 cm
Juvenile
photo: R. CLARKE & J. L. MCKAY

RT06251-4 *Myuchelys purvisi* (Wells & Wellington, 1985)
Manning River, NSW, Australia
18–22 cm

photo: J. Cann

35°C 3

RT06252-4 *Myuchelys purvisi* (Wells & Wellington, 1985)
Manning River, NSW, Australia
18–22 cm

photo: J. Cann

35°C 3

RT06261-4 *Rheodytes leukops* LEGLER & CANN, 1980
Connors River, QLD, Australia
24–27 cm

photo: J. CANN

35°C 3

RT06262-4 *Rheodytes leukops* LEGLER & CANN, 1980
Connors River, QLD, Australia
24–27 cm

photo: A. ELLIOTT

35°C 3

RT06263-4 *Rheodytes leukops* LEGLER & CANN, 1980
Connors River, QLD, Australia
24–27 cm

photo: J. CANN

35°C 3

RT06264-4 *Rheodytes leukops* LEGLER & CANN, 1980
Connors River, QLD, Australia
24–27 cm

photo: A. ELLIOTT

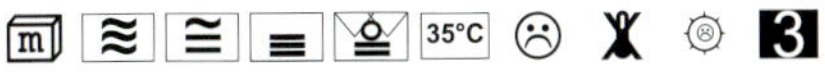

RT06265-4 *Rheodytes leukops* Legler & Cann, 1980
Connors River, QLD, Australia
24–27 cm

photo: A. Elliott

Rheodytes leukops (red)

RT06266-1 *Rheodytes leukops* Legler & Cann, 1980
Connors River, QLD, Australia
24–27 cm
Hatchling

photo: J. Cann

RT06271-4 *Pseudemydura umbrina* SIEBENROCK, 1901
Ellenbrook Nature Reserve, Perth Region, WA, Australia
14–16 cm

photo: J. LOCHMAN/LOCHMAN LT

RT06272-4 *Pseudemydura umbrina* SIEBENROCK, 1901
Perth Region, WA, Australia
14–16 cm

photo: G. KUCHLING

RT06273-4 *Pseudemydura umbrina* SIEBENROCK, 1901
Perth Region, WA, Australia
14–16 cm

photo: G. KUCHLING

RT06274-4 *Pseudemydura umbrina* SIEBENROCK, 1901
Captive specimen
14–16 cm

photo: R. MIDTGAARD

35°C 3

Pseudemydura umbrina (red)

RT06275-1 *Pseudemydura umbrina* SIEBENROCK, 1901
Captive bred, Perth Zoo, WA, Australia
14–16 cm
Hatching

photo: J. LOCHMAN/LOCHMAN LT

35°C 3

RT06276-1 *Pseudemydura umbrina* SIEBENROCK, 1901

14–16 cm
Hatchling

photo: C. LATTA

35°C 3

RT06277-4 *Pseudemydura umbrina* SIEBENROCK, 1901
Perth Region, WA, Australia
14–16 cm

photo: G. KUCHLING

35°C 3

RT06278-4 *Pseudemydura umbrina* Siebenrock, 1901
Ellenbrook Nature Reserve, Perth Region, WA, Australia
14–16 cm

photo: J. Lochman/Lochman LT

35°C 3

RT06279-4 *Pseudemydura umbrina* Siebenrock, 1901

14–16 cm

photo: C. Latta

35°C 3

RT06280-4 *Pseudemydura umbrina* Siebenrock, 1901
Bullsbrook, Perth, WA, Australia
14–16 cm

photo: G. Schmida

35°C 3

Appendix

Taxa that were described or revalidated after the publication of vols. 1–4 respectively.
Taxa, die nach der Veröffentlichung der Bände 1–4 beschrieben bzw. revalidiert wurden.

Symbols used in the table below / *Verwendete Symbole in der folgenden Liste:*
► = at least one photo of this taxon is shown in the appendix; *mindestens ein Foto dieses Taxons wird gezeigt.*
► = a reference is provided where a photo of this taxon has been shown in previous volumes of the series / *es folgt ein Verweis auf Fotos, die in anderen Bänden bereits gezeigt wurden.*

Macrochelys suwanniensis Thomas, Granatovsky, Bourque, Krysko, Moler, Gamble, Suarez, Leone, Enge & Roman, 2014 ►

Kinosternon vogti López-Luna, Cupul-Magaña, Escobedo-Galván, González-Hernández, Centenero-Alcalá, Rangel-Mendoza, Ramírez-Ramírez & Cazares-Hernández, 2018 ►

Sternotherus intermedius Scott, Glenn & Rissler, 2018 ►

Graptemys pearlensis Ennen, Lovich, Kreiser, Selman & Qualls, 2010 ► ***Vol. 2***, p. 53 no. RT01650-4, RT01651-4

Trachemys grayi panamensis McCord, Joseph-Ouni, Hagen & Blanck, 2010 ►

Trachemys medemi Vargas-Ramírez, del Valle, Ceballos & Fritz, 2017 ►

Trachemys venusta iversoni McCord, Joseph-Ouni, Hagen & Blanck, 2010 ► ***Vol. 3***, p. 41 no. RT02820-4, RT02821-4, RT02822-4, RT02823-2, RT02824

Trachemys venusta uhrigi McCord, Joseph-Ouni, Hagen & Blanck, 2010 ► ***Vol. 3***, p. 41 no. RT02810-4

Batagur affinis affinis (Cantor, 1847) ► ***Vol. 4***, p. 29 no. RT04011-4, RT04013-1, RT04015

Batagur affinis edwardmolli Praschag, Holloway, Georges, Päckert, Hundsdörfer & Fritz, 2009 ►

Batagur baska (Gray, 1831) ►
[All photos present in volume 4 need to be attributed to *Batagur affinis affinis* according to the latest systematics / *Alle Fotos, die in Band 4* Batagur baska *zugeordnet wurden, zeigen neuesten Erkenntnissen zufolge* Batagur affinis affinis].

Cuora aurocapitata dabieshani Blanck, Protiva, Zhou, Li, Crow & Tiedemann, 2017 ►

Cuora cyclornata cyclornata Blanck, McCord & Le, 2006 ►

Cuora cyclornata annamitica Blanck, Protiva, Zhou, Li, Crow & Tiedemann, 2017 ►

Cuora cyclornata meieri Blanck, McCord & Le, 2006 ► ***Vol. 4***, p.97 no. RT04841-4, RT04842-4

RT07072-4 *Batagur baska*
♂ in breeding colouration

photo: P. Praschag

RT07073-4 *Batagur baska*
♀

photo: P. Praschag

RT07074-4 *Batagur baska* – Juvenile (First ever captive-bred), Tiergarten Schönbrunn, Vienna, Austria

photo: P. Praschag

RT07011-4 *Macrochelys suwanniensis*
Suwannee River, Florida, USA

photo: R. T. Zappalorti

RT07021-4 *Kinosternon vogti*
Puerto Vallarta, Jalisco, Mexico

photo: M. A. Lopez-Luna

RT07022-4 *Kinosternon vogti*
Puerto Vallarta, Jalisco, Mexico

photo: M. A. Lopez-Luna

RT07023-4 *Kinosternon vogti*
Puerto Vallarta, Jalisco, Mexico

photo: M. A. Lopez-Luna

RT07031-4 *Sternotherus intermedius*
NW Florida, USA

photo: P. A. Scott

RT07041-4 *Trachemys grayi panamensis*
Panama

photo: C. Hagen

RTH0003-4 *Trachemys medemi*
Río Atrato, Colombia

photo: C. del Valle

RTH0003-4 *Trachemys medemi*
Río Atrato, Colombia

photo: C. del Valle

RT07061-4 *Batagur affinis edwardmolli*
Cambodia

photo: Brian D. Horne

RT07062-4 *Batagur affinis edwardmolli*
Cambodia

photo: Brian D. Horne

RT07081-4 *Cuora aurocapitata dabieshani*
Anhui, China

photo: T. Blanck

RT07082-4 *Cuora aurocapitata dabieshani*
Anhui, China

photo: T. Blanck

Cuora trifasciata luteocephala Blanck, Protiva, Zhou, Li, Crow & Tiedemann, 2017 ▶ ***Vol. 4***, p. 97 no. RT04821-4, RT04822-4

Cyclemys enigmatica Fritz, Guicking, Auer, Sommer, Wink & Hundsdörfer 2008 ▶

Cyclemys fusca Fritz, Guicking, Auer, Sommer, Wink & Hundsdörfer 2008 ▶ ***Vol. 4***, p. 55 no. RT04361-4, RT04362-4

Cyclemys gemeli Fritz, Guicking, Auer, Sommer, Wink & Hundsdörfer 2008 ▶ ***Vol. 4***, p. 55 no. RT04331-4

Malayemys khoratensis Ihlow, Vamberger, Flecks, Hartmann, Cota, Makchai, Meewattana, Dawson, Kheng, Rödder & Fritz, 2016 ▶

Chelonoidis chathamensis (Van Denburgh, 1907) ▶ ***Vol. 3***, p. 90 no. RT03450-4

Chelonoidis donfaustoi Poulakakis, Edwards & Caccone *in* Poulakakis, Edwards, Chiari, Garrick, Russello, Benavides, Watkins-Colwell, Glaberman, Tapia, Gibbs, Cayot & Caccone, 2015 ▶

Chelonoidis guntheri (Baur, 1889) ▶ ***Vol. 3***, p. 90 no. RT03440-4, RT03441-4, RT03442-4

Gopherus evgoodei Edwards, Karl, Vaughn, Rosen, Meléndez Torres & Murphy, 2016 ▶ ***Vol. 2***, p. 117 no. RT02330-4

Gopherus morafkai Murphy, Berry, Edwards, Leviton, Lathrop & Riedle, 2011 ▶ ***Vol. 2***, p. 117 no. RT02340-4, RT02341-4

Lissemys ceylonensis (Gray, 1856) ▶ ***Vol. 4***, p. 136–137 no. RT05221-4, RT05222-4, RT05223-4

Lissemys punctata vittata (Peters, 1854) ▶ ***Vol. 4***, p. 135 no. RT05203-4

Amyda cartilaginea maculosa Fritz, Gemel, Kehlmaier, Vamberger & Praschag, 2014 ▶ ***Vol. 4,*** p. 124 no. RT05083-4

Amyda ornata ornata (Gray, 1861) ▶ ***Vol. 4***, p. 124–125 no. RT05092-4, RT05093-4, RT05096-2

Amyda ornata phayrei (Theobald, 1868) ▶ ***Vol. 4***, p. 125 no. RT05094-2

Pelodiscus axenaria (Zhou, Zhang & Fang, 1991) ▶

Pelodiscus parviformis Tang, 1997 ▶ ***Vol. 4***, p. 144–145 no. RT05301-4, RT05302-4, RT05303-4, RT05304-4, RT05305-4

Mesoclemmys perplexa Bour & Zaher, 2005 ▶

Chelodina (Chelodina) mccordi roteensis McCord, Joseph-Ouni & Hagen, 2007 ▶

Chelodina (Chelodina) mccordi timorensis McCord, Joseph-Ouni & Hagen, 2007 ▶

Pelomedusa barbata Petzold, Vargas-Ramirez, Kehlmaier, Vamberger, Branch, Du Preez, Hofmeyr, Meyer, Schleicher, Široky & Fritz, 2014 ▶ ***Vol. 1*** *(2 ed.)*, p. 130 no. RT01115-4

Pelomedusa galeata (Schoepff, 1792) ▶ ***Vol. 1*** *(2 ed.)* p. 131 no. RT01121-4

Pelomedusa gehafie (Rüppell, 1835) ▶

Pelomedusa kobe Petzold, Vargas-Ramirez, Kehlmaier, Vamberger, Branch, Du Preez, Hofmeyr, Meyer, Schleicher, Široky & Fritz, 2014 ▶

Pelomedusa neumanni Petzold, Vargas-Ramirez, Kehlmaier, Vamberger, Branch, Du Preez, Hofmeyr, Meyer, Schleicher, Široky & Fritz, 2014 ▶ ***Vol. 1*** *(2 ed.)* p. 130 no. RT01111-4, RT01113-4

Pelomedusa olivacea (Schweigger, 1812) ▶

Pelomedusa schweinfurthi Petzold, Vargas-Ramirez, Kehlmaier, Vamberger, Branch, Du Preez, Hofmeyr, Meyer, Schleicher, Široky & Fritz, 2014 ▶

Pelomedusa somalica Petzold, Vargas-Ramirez, Kehlmaier, Vamberger, Branch, Du Preez, Hofmeyr, Meyer, Schleicher, Široky & Fritz, 2014 ▶ *Vol. 1*, p. 131 no. RT01116-4, RT01117-4

Pelomedusa variabilis Petzold, Vargas-Ramirez, Kehlmaier, Vamberger, Branch, Du Preez, Hofmeyr, Meyer, Schleicher, Široky & Fritz, 2014 ▶

RT07091-4 *Cuora cyclornata cyclornata*
Central Vietnam

photo: T. Blanck

RT07092-4 *Cuora cyclornata cyclornata*
Central Vietnam

photo: T. Blanck

RT07101-4 *Cuora cyclornata annamitica*
South Central Vietnam

photo: T. Blanck

RT07102-4 *Cuora cyclornata annamitica*
South Central Vietnam

photo: T. Blanck

RT07111-4 *Cyclemys enigmatica*
Rantau Prapat, Sumatra, Indonesia

photo: M. Gaulke

RT07121-4 *Malayemys khoratensis*
Sikhio, Thailand

photo: F. Ihlow

RT07122-4 *Malayemys khoratensis*
Udon Thani, Thailand

photo: F. Ihlow

RT07131-4 *Chelonoidis donfaustoi*
Cerro Fatal, Santa Cruz,
Galápagos, Ecuador

photo: P.P. van Dijk

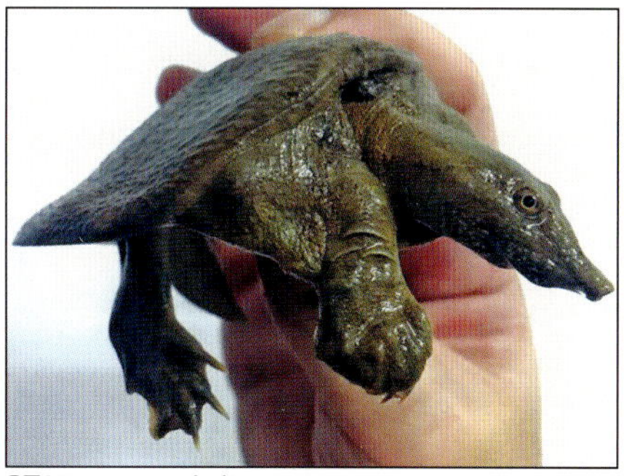
RT07141-4 *Pelodiscus axenaria*
Captive specimen

photo: Chan Tai Man

RT07142-4 *Pelodiscus axenaria*
Captive specimen

photo: Chan Tai Man

RT07151-4 *Mesoclemmys perplexa*
Piauí, Brazil

photo: V. Carvalho courtesy of R.C. Vogt

RT07152-4 *Mesoclemmys perplexa*
Piauí, Brazil

photo: V. Carvalho courtesy of R.C. Vogt

RT07161-4 *Chelodina (Ch.) mccordi roteensis*
Eastern Rote Island, Indoesia

photo: C. Hagen

RT07171-4 *Chelodina (Ch.) mccordi timorensis*
Lautém District, East Timor, Indonesia

photo: M. O'Shea

RT07174-4 *Chelodina (Ch.) mccordi timorensis*
Lautém District, East Timor, Indonesia

photo: M. O'Shea

RT07172-4 *Chelodina (Ch.) mccordi timorensis*
Lautém District, East Timor, Indonesia

photo: C. Hagen

RT07182-4 *Pelomedusa gehafie*
Ghinda, Eritrea

photo: T. Mazuch

RT07181-4 *Pelomedusa gehafie*
Ghinda, Eritrea

photo: T. Mazuch

RT07191-2 *Pelomedusa kobe*
Naberera, Manyara, Tanzania
ZSM 334/1978:1, Juvenile

photo: E. Morawa

RT07201-4 *Pelomedusa olivacea*
Near Niakar, Senegal

photo: T. Diagne/African Chelonian Institute

RT07202-4 *Pelomedusa olivacea*
Senegal
♂

photo: T. Diagne/African Chelonian Institute

RT07211-4 *Pelomedusa schweinfurthi*
Liria, Central Equatoria, South Sudan
♀, SMF 56161

photo: P. Široký & E. Morawa

RT07221-4 *Pelomedusa variabilis*
Grand Popo, Benin

photo: T. Diagne/African Chelonian Institute

RT07222-4 *Pelomedusa variabilis*
Togo

photo: T. Diagne/African Chelonian Institute

Symbols/*Symbole*

In order to depict as many turtle species and subspecies as possible, and to reckon with the worldwide distribution of our picture volumes, we have decided to do without exhaustive legends, replacing them by international understandable pictograms from which any reader may conclude the basic features and demands of the individual species in husbandry.
Um möglichst viele Schildkröten im Bild zeigen zu können, und um dem weltweiten Vertrieb unserer Bildbände Rechnung zu tragen, haben wir bewusst auf ausführliche Texte verzichtet und ersetzen diese durch international verständliche Symbole, mit deren Hilfe jeder Leser die wichtigsten Eigenschaften und Bedürfnisse der verschiedenen Arten in Menschenobhut erkennen kann.

Codenumber/*Codenummer*
The first two numbers denote the animal group; "R" is for "Reptiles, "T" for "Turtles"; each taxon acquires hereby (together with the following five numbers) its distinct codenumber. The codenumbers allow international communication. The last figure of the codenumber depicts the age of the animal shown.
Die ersten zwei Buchstaben erläutern die Tiergruppe; "R" steht für "Reptiles", "T" für "Turtles"; zusammen mit den fünf folgenden Zahlen erhält jedes Taxon so seine unverwechselbare Codenummer. Die Codenummern dienen der Verständigung auf internationaler Ebene. Die letzte Ziffer verweist auf das Alter des abgebildeten Tieres:
1 = hatchling/*Schlüpfling*
2 = juvenile/*Jungtier*
3 = subadult/*subadult*
4 = adult/*adult*
5 = very old specimen/*sehr altes Exemplar*

Sex/*Geschlecht*
♂ male/*männlich*, ♀ female/*weiblich*, ♂♀ couple/*Paar*

Size/*Größe*
in cm = the approx. straight carapace length the turtles can achieve as adults/
in cm = ungefähre Carapaxlänge (Stockmaß; nicht über die Panzerwölbung gemessen), die die Schildkröten als Adulti erreichen können

Terrarium Size/*Terrariengröße*
Minimum size of terrarium for 1–2 turtles (CL = Straight carapace length); add 20 % of space for each additional specimen/Mindestgröße Terrarium für 1–2 Schildkröten (CL = Stockmaß Carapaxlänge); für jedes weitere Exemplar 20 % Grundfläche zusätzlich

S	Length/*Länge* = CL × 4	Width/*Breite* = CL × 2
M	Length/*Länge* = CL × 6	Width/*Breite* = CL × 3
L	Length/*Länge* = CL × 8	Width/*Breite* = CL × 4
XL	Length/*Länge* = CL × 20	Width/*Breite* = CL × 10

Water Level/*Wasserstand*
medium, shell width × 3/*mittel, Panzerbreite × 3*
high, shell width × 5/*hoch, Panzerbreite × 5*

Settings/*Behältereinrichtung*
arid: sand, earth, stones, roots, water dish/
trocken: Sand, Erde, Steine, Wurzeln, Trinkgefäß
semi-humid: sphagnum, earth, dead leaves, stones, roots, large water basin (land section 70 %, *water section 30 %; water level: carapace height × 0,5)/*
halbfeucht: Torfmoos, Erde, Laub, Steine, Wurzeln, großes Wasserbecken (Landteil 70 %, Wasserteil 30 %; Wasserstand Panzerhöhe × 0,5)
humid: sphagnum, earth, dead leaves, stones, roots, very large water basin (land section 50 %, water section 50 %; water level: shell width)/*feucht: Torfmoos, Erde, Laub, Steine, Wurzeln, sehr großes Wasserbecken (Landteil 50 %, Wasserteil 50 %; Wasserstand Panzerbreite)*
water basin with stones, roots and plastic plants as submersed hiding places/*Wasserteil mit Steinen, Wurzeln und Plastikpflanzen als Unterwasserverstecke*
high layer of fine sand in the water basin/
hohe feinkörnige Sandschicht im Wasserteil
thin layer of gravel or coarse sand in the water basin/
dünne Kies- oder grobkörnige Sandschicht im Wasserteil
thin layer of fine sand in the water basin/
dünne feinkörnige Sandschicht im Wasserteil
land section mainly for oviposition/
Landteil hauptsächlich zur Eiablage
land section for oviposition and basking/
Landteil für Eiablage und Sonnenbäder
H specialist: caves as hiding places/
Spezialist: Höhlen als Unterschlupf
S specialist: rock crevices as hiding places/
Spezialist: Felsspalten als Unterschlupf
<Bw> specialist: brackish water dweller/
Spezialist: Brackwasserbewohner
<Sw> specialist: saltwater dweller/
Spezialist: Salzwasserbewohner

Temperature (at least 23–26 °C at daytime)/
Temperatur (tagsüber mindestens 23–26 °C)
no additional radiation heat required/
keine zusätzliche Strahlungswärme erforderlich
35°C radiation heat required locally up to approx. 35 °C/
Strahlungswärme lokal bis ca. 35 °C
45°C radiation heat required locally up to approx. 45 °C/
Strahlungswärme lokal bis ca. 45 °C

Diet/*Ernährung*
mainly carnivorous/*überwiegend Fleischfresser*
mainly herbivorous/*überwiegend Pflanzenfresser*
omnivorous/*Allesfresser*

Periods of Dormancy/*Ruheperioden*
hibernation at 4–8 °C/*Winterruhe bei 4–8 °C*
in winter, reduced activity at lower temperatures/
im Winter reduzierte Aktivität bei abgesenkten Temperaturen
no hibernation/*keine Winterruhe*
aestivation/*Sommerruhe*

Outdoor husbandry/*Freilandhaltung*
recommended/*empfohlen*
possible at high temperatures/*bei hohen Temperaturen möglich*
not recommended/*nicht empfohlen*

Group composition/*Vergesellschaftung*
0 separation of sexes recommended/
Geschlechtertrennung empfohlen
1 individual housing of at least the males recommended/
Einzelhaltung zumindest der Männchen empfohlen
2 ♂♀
3 ♂♀♀
4 ♂♂♀♀

Peculiarities/*Besonderheiten*
suitable for beginners as well/auch für Anfänger geeignet
attention: suitable for experienced keepers only/
Vorsicht, nur für erfahrene Halter geeignet
protected species (appendices CITES I–III and/or EU A–C)/*geschützte Art (Anhänge CITES I–III und/oder EU A–C)*

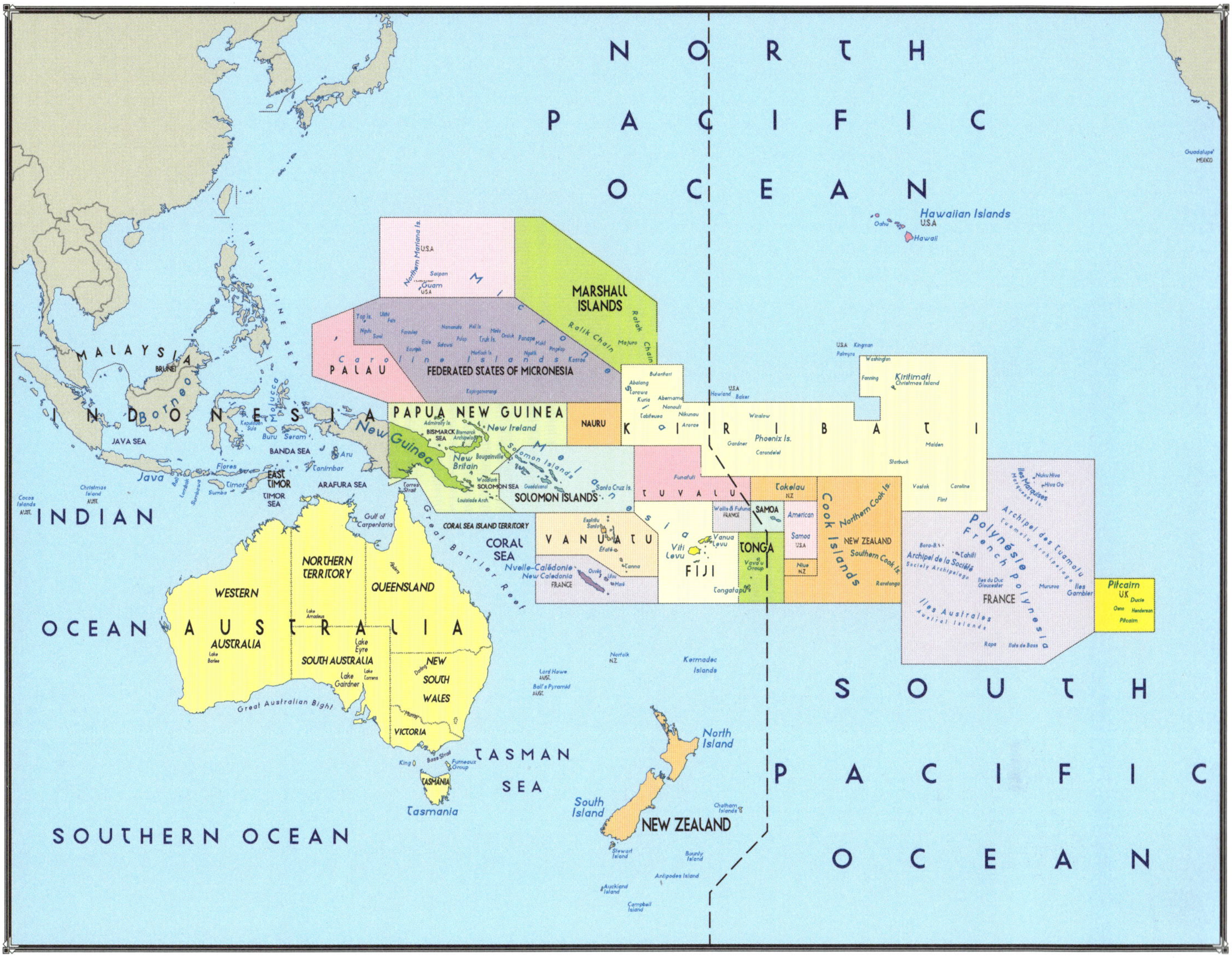
NORTH PACIFIC OCEAN
SOUTH PACIFIC OCEAN
INDIAN OCEAN
SOUTHERN OCEAN
TASMAN SEA
CORAL SEA
JAVA SEA
BANDA SEA
ARAFURA SEA
TIMOR SEA
BISMARCK SEA
SOLOMON SEA
Hawaiian Islands
U.S.A.
Oahu
Hawaii
MARSHALL ISLANDS
Ralik Chain
Ratak Chain
Majuro
Northern Mariana Is.
U.S.A.
Saipan
Guam
U.S.A.
Micronesia
PALAU
Caroline Islands
FEDERATED STATES OF MICRONESIA
Yap Is.
Ulithi
Faraulep
Namonuito
Hall Is.
Truk Is.
Ponape
Mokil
Pingelap
Kosrae
Kapingamarangi
PAPUA NEW GUINEA
New Guinea
New Ireland
New Britain
Bougainville
Admiralty Is.
Bismarck Archipelago
Woodlark
Louisiade Arch.
NAURU
KIRIBATI
Phoenix Is.
Kiritimati
Christmas Island
Washington
Fanning
Palmyra
Kingman
U.S.A.
Howland
Baker
Winslow
Gardner
Canton
Malden
Starbuck
Vostok
Caroline
Flint
Banaba
Tarawa
Abaiang
Abemama
Nonouti
Nikunau
Arorae
Butaritari
Solomon Islands
SOLOMON ISLANDS
Guadalcanal
Santa Cruz Is.
Melanesia
TUVALU
Funafuti
Tokelau
N.Z.
Wallis & Futuna
FRANCE
SAMOA
American Samoa
U.S.A.
VANUATU
Espiritu Santo
Efaté
Tanna
Nvelle-Calédonie
New Caledonia
FRANCE
FIJI
Viti Levu
Vanua Levu
TONGA
Vava'u Group
Tongatapu
Niue
N.Z.
Cook Islands
Northern Cook Is.
Southern Cook Is.
NEW ZEALAND
Rarotonga
Polynésie Française
French Polynesia
FRANCE
Archipel de la Société
Society Archipelago
Tahiti
Bora-Bora
Archipel des Tuamotu
Tuamotu Archipelago
Îles Marquises
Marquesas Is.
Nuku Hiva
Hiva Oa
Îles Gambier
Îles Australes
Austral Islands
Rapa
Îlots de Bass
Mururoa
Pitcairn
U.K.
Ducie
Oeno
Henderson
Pitcairn
CORAL SEA ISLAND TERRITORY
Great Barrier Reef
MALAYSIA
BRUNEI
Borneo
INDONESIA
Java
Bali
Lombok
Sumbawa
Flores
Sumba
Timor
EAST TIMOR
Buru
Seram
Aru
Tanimbar
Moluccas
PHILIPPINE SEA
Christmas Island
AUST.
Cocos Islands
AUST.
AUSTRALIA
WESTERN AUSTRALIA
NORTHERN TERRITORY
QUEENSLAND
SOUTH AUSTRALIA
NEW SOUTH WALES
VICTORIA
TASMANIA
Tasmania
Gulf of Carpentaria
Great Australian Bight
Lake Eyre
Lake Gairdner
Lake Torrens
Lake Amadeus
Lake Barlee
King
Furneaux Group
Bass Strait
Torres Strait
Lord Howe
AUST.
Ball's Pyramid
AUST.
Norfolk
N.Z.
Kermadec Islands
NEW ZEALAND
North Island
South Island
Stewart Island
Chatham Islands
Bounty Island
Antipodes Island
Auckland Island
Campbell Island
Guadalupe
MEXICO

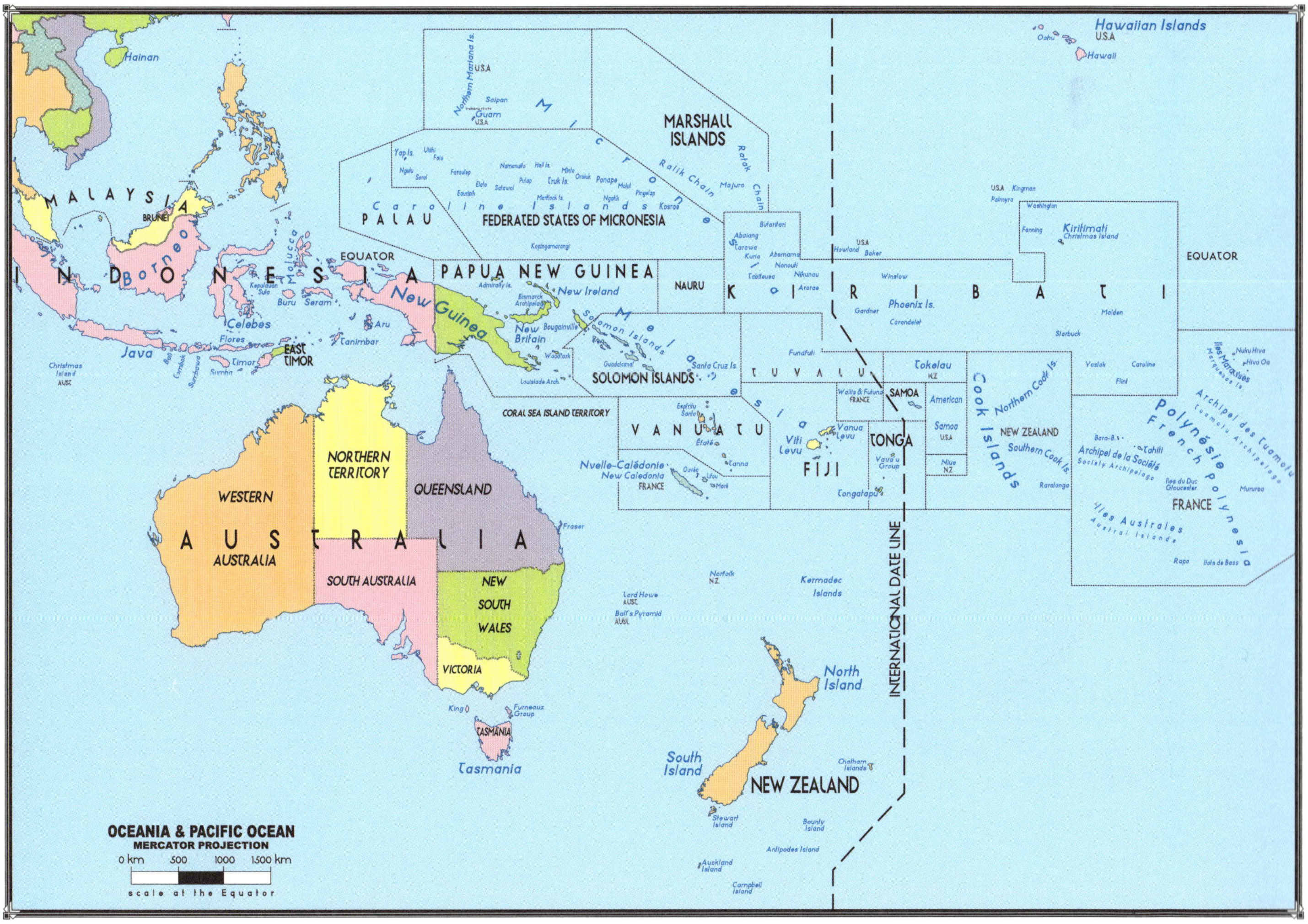
Hainan
Hawaiian Islands
U.S.A.
Oahu
Hawaii
Northern Mariana Is.
U.S.A.
Saipan
Guam
U.S.A.
Micronesia
MARSHALL ISLANDS
Ralik Chain
Ratak Chain
Majuro
PALAU
Caroline Islands
FEDERATED STATES OF MICRONESIA
Kapingamarangi
MALAYSIA
BRUNEI
Borneo
INDONESIA
Celebes
Java
Flores
Timor
Sumba
EAST TIMOR
Buru
Seram
Aru
Tanimbar
Moluccas
EQUATOR
PAPUA NEW GUINEA
New Guinea
New Ireland
New Britain
Bougainville
Bismarck Archipelago
Louisiade Arch.
NAURU
KIRIBATI
Phoenix Is.
Kiritimati
Christmas Island
Washington
Fanning
Palmyra
Kingman
U.S.A.
Howland
Baker
Winslow
Gardner
Malden
Starbuck
Vostok
Caroline
Flint
EQUATOR
Solomon Islands
SOLOMON ISLANDS
Santa Cruz Is.
Melanesia
TUVALU
Funafuti
Tokelau
N.Z.
Wallis & Futuna
FRANCE
SAMOA
American Samoa
U.S.A.
CORAL SEA ISLAND TERRITORY
VANUATU
Espiritu Santo
Efaté
Tanna
Nvelle-Calédonie
New Caledonia
FRANCE
FIJI
Viti Levu
Vanua Levu
TONGA
Vava'u Group
Tongatapu
Niue
N.Z.
Cook Islands
Northern Cook Is.
Southern Cook Is.
NEW ZEALAND
Rarotonga
Polynésie Française
French Polynesia
FRANCE
Archipel de la Société
Society Archipelago
Tahiti
Bora-Bora
Archipel des Tuamotu
Tuamotu Archipelago
Îles Marquises
Marquesas Is.
Nuku Hiva
Hiva Oa
Mururoa
Îles Australes
Austral Islands
Rapa
Îlots de Bass
Christmas Island
AUST.
AUSTRALIA
WESTERN AUSTRALIA
NORTHERN TERRITORY
QUEENSLAND
SOUTH AUSTRALIA
NEW SOUTH WALES
VICTORIA
TASMANIA
Tasmania
Fraser
King
Furneaux Group
Lord Howe
AUST.
Ball's Pyramid
AUST.
Norfolk
N.Z.
Kermadec Islands
INTERNATIONAL DATE LINE
NEW ZEALAND
North Island
South Island
Stewart Island
Chatham Islands
Bounty Island
Antipodes Island
Auckland Island
Campbell Island
OCEANIA & PACIFIC OCEAN
MERCATOR PROJECTION
0 km 500 1000 1500 km
scale at the Equator